Boatmen's Guide to Light Salvage

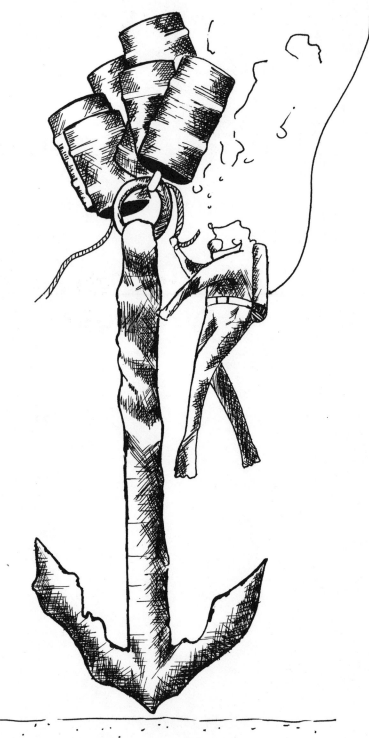

Salving a 17th century anchor.

Boatmen's Guide to Light Salvage

GEORGE H. REID

CORNELL MARITIME PRESS, INC.
Centreville   1979   Maryland

Copyright © 1979 by Cornell Maritime Press, Inc.

All Rights Reserved

No part of this book may be used or reproduced in any manner whatsoever without written permission except in the case of brief quotations embodied in critical articles and reviews. For information address Cornell Maritime Press, Inc., Centreville, Maryland 21617

Library of Congress Cataloging in Publication Data

Reid, George H 1924-
   Boatmen's guide to light salvage.

   Includes index.
   1. Salvage.   2. Boats and boating.   I. Title.
VK1491.R44   627.7'03   78-27593
ISBN 0—87033—248-1

Printed and Bound in the United States of America

Contents

		Page
Preface		vii
1.	An Introduction to Light Salvage	1
2.	Basic Skills Required	4
3.	The Gear	6
4.	Common Salvage Situations	10
5.	Refloating Grounded Vessels	12
6.	Dewatering the Sunken or Sinking Vessel	17
7.	Raising Submerged Vessels	23
8.	Vessel Adrift — Rescue Towing	31
9.	Submerged and Lost	33
10.	Seamanship, Rigging and Beach Gear	38
11.	Rigging for a Tow	52
12.	Moving Material Under Water	57
13.	Miscellaneous and Extraneous	61
14.	Some Legal Aspects of Salvage	64
15.	Getting Started in Light Salvage	67
Index		69

Dedicated
to

JOHN HOOPER, SAILOR

Preface

When I was a youngster, my next door neighbor, a very kind middle-aged man, set out to teach me about the sea. He had served in sail and knew just about everything there was to know about boats.

Having a tutor like this was a tremendous help, and before we moved away and lost contact, I had learned to row and scull and sail a boat and acquired quite a few of the sailorly skills that would stand me in good "stead" when I eventually went to sea myself.

The last time I saw him he was 83 years old and still quite spry. I had my Third Mates ticket by that time and he was as pleased as I.

The lessons he taught me many years ago are just as important today. And, of those who are not fortunate enough to have such a mentor, some will perhaps learn their lessons through bitter experience.

The intent of this volume is to provide the same sort of basic information about a slightly different aspect of seamanship that my old friend so generously provided for me. For before one can help save another's boat, he must first of all be able to save his own.

George H. Reid

Also by George H. Reid

A Primer of Towing

CHAPTER 1

An Introduction to Light Salvage

Salvage is a romantic word. It stimulates the imagination and conjures visions of rescue tugs driving through mountainous seas en route to aid a stricken vessel. At any rate, this type of operation receives the attention of the press and reaches the public eye. This view has led to a general tendency to regard all salvage operations as actions of heroic proportions, requiring skills so arcane that they are far beyond the comprehension of ordinary mortals who seek their pleasure or livelihood from the sea.

This popular concept is not altogether accurate. The act of patching and refloating a sunken skiff or towing in a disabled yacht is categorically as much an act of salvage as refloating a stranded supertanker, although admittedly there is a lot more work involved in the latter. There are plenty of small salvage operations being carried out all the time, almost unnoticed except by those involved.

I do not wish to destroy a myth, but salvage work mainly consists of carrying on commonplace activities under unusual circumstances. There are many "tricks of the trade" in this line of work (as in most others) and salvors are just as reluctant to disclose them as those in other employment. Salvage is just good seamanship coupled with a few other skills and applied with imagination.

My reason for dealing with this topic is twofold: first, to provide a source of information for the pleasure boatman on some aspects of seamanship not ordinarily dealt with in most books on this subject; second, to encourage the seasoned boatman and/or diver to expand his activities to include the possibility of engaging in salvage for commercial purposes. This might be done as much for fun as profit, but at least it will probably defray some of the costs of his equipment.

I feel there is justification in both instances. The number of pleasure boats is increasing every year, and more of these vessels are venturing offshore than ever before. The boatman should be prepared to cope with emergencies for his own security and be capable of rendering assistance to others in peril. The age-old "Law of the Sea" applies here. In all too many instances, boatmen have come to rely on an overworked Coast Guard for assistance, when with a little effort and a few additional items of equipment aboard, they would be able to fend for themselves or assist others in an emergency.

The experienced boatman and diver, especially one who is mechanically inclined and proficient with hand tools may be able to use the information contained here towards a commercial end. The increased traffic on our waterways has resulted in increased casualties. The capable salvage man is the obvious beneficiary of this situation. Compensation to the successful salvor is usually generous and marine insurance companies deliberately encourage them since it is far better to recover any part of the insured equipment than suffer a total loss.

There is a tendency to consider the term "light salvage" as applicable only when small craft are involved. This is not necessarily the case. To me it is indicated more by what is required in effort and equipment to do the job, than the size of the vessel or object to be salvaged.

For example, a barge 120' x 30' x 7' raised with a 2-inch gasoline pump, towed to its berth by a 16' boat propelled by a 40 hp outboard motor is a "light salvage" operation. An acquaintance of mine subcontracted to raise a tank barge of 1,200 tons burden that had sunk near shore. He built a raft of oil drums for use as a working platform, fitting it out with a beach umbrella and ice chest. The barge was raised on air from a compressor on the nearby beach. By systematically plugging all the leaks in the decks and topsides (easily done as bubbles showed the locations), the barge was raised. When the barge was afloat, a bulldozer dragged it ashore bit by bit, while one of the local junk dealers cut it into scrap with acetylene torches. For approximately three weeks of hard work the salvor made almost $10,000.00. This, too, was "light salvage."

I once ran 40 miles in a 17' Boston whaler to an empty 200' tank barge that had gone ashore on a reef. Diving the area indicated that the barge had struck head-on and then drifted into a broadside position. Since this was the case, it couldn't have been too hard aground. This was confirmed when at high tide the stern would lift about ½" from the bottom as waves passed. An examination of the hull showed negligible damage. The surrounding area was buoyed with large plastic capped bottles to indicate safe depths and a tug was called to assist. The barge came off at the next high tide. The fee for the tug was $2,400, and the salvage fee was considerably more than that. This was a "light salvage" operation except for the assisting tug which was paid for by the owner of the barge. My own costs were for a helper, the gas and oil for the whaler, and two tanks of air!

I recall witnessing the recovery of an anchor with three shots of chain lost by a passenger ship in St. Thomas harbor. The salvor located and attached a cable to the bitter end of the chain, and then buoyed the end of the cable. When the ship returned to port a deal was made with the captain. Upon sailing, the vessel maneuvered close to the buoy and passed a wire pennant through the hawsepipe which was connected to the cable. Both chain and anchor were soon heaved aboard ship. Although I never learned exactly how much the salvor was paid, it must have been a fair amount as he proceeded to throw a party which lasted about three weeks.

I do feel that there are several conditions which must be met for a job to qualify as a "light marine salvage operation."

1. If diving is involved, depth should be restricted to 60' or less due to the limited working times at greater depths and additional hazards involved.

2. The amount of diving required must be of such nature that SCUBA gear is adequate.

3. No exotic techniques should be required. Underwater burning, welding and explosives should be left to the experts.

4. All gear required for the operation should fit into a husky small boat (17' to 30') or on a work float that is easily towed about by such a boat—*unless* the job site is close enough to a beach or wharf to enable hoses, etc., to be extended out to the wreck.

5. The location and nature of operations should permit the salvors to work with minimum risks to themselves and their equipment.

If these conditions are met and the salvor is a good diver and/or boatman (preferably both), he is then in a position to undertake a salvage with a fair likelihood of success. As evidence of this, there are a good number of people who either supplement their income or earn most of their livelihood from engaging in this activity.

Chapter 2

Basic Skills Required

In many instances salvage work is like rendering first aid. The salvor is obliged to make hasty but effective repairs that will serve until permanent repairs can be carried out. He frequently must improvise and make patches out of the material at hand.

The salvor has to be adaptable in other ways and must know how to utilize a situation to his own best advantage. Mechanical ability is a definite asset as pumps and compressors are often necessary and familiarity with their operation is important. He must also be able to carry out his operation under rather trying conditions. In order to do so with minimum risk to himself, his helpers, and his equipment, he should have the abilities listed below, more or less in the order shown:

1. *Swimming.* Even if the salvor is not a diver, he should be a good swimmer. Obviously, anyone having anything to do with boats should be a swimmer.

2. *Boatmanship.* The salvor should know how to operate and control his boat, with or without a tow, under a variety of conditions. Salvages are often carried out in exposed areas, and lack of this skill cannot only botch the job, but result in the loss of equipment and injury to personnel.

3. *Seamanship.* It is essential to be able to make lines fast both above and under water so they will hold, and in such a fashion that they can later be cast off even after the line has swollen or the knot has pulled tight from a heavy strain. He should know how to use tide tables, be knowledgeable about wind and current, and be able to do the piloting or navigation necessary to get his workboat safely to and from its destination. It is important to be familiar with using ground tackle, taking soundings and taking marks for relocating a wreck without buoying it. He should know how to rig his boat for towing, use chafing gear, and throw a heaving line. He should also know how to pass a stopper on wire cable or fiber hawser, and to make the standard knots and splices, lash things securely and make slings for handling variously shaped objects. Good salvage work requires a sound knowledge of seamanship and the ability to employ it with imagination. Much of the information needed will be supplied in Chapter 10.

4. *Rigging.* The salvor must also be a pretty fair rigger. If you have ever been to boat shows just before they open and have seen the rigging gangs unloading and moving those big yachts into position, you know exactly what I'm talking about. Salvage rigging is a bit different, but employs the same type of approach. The salvor uses anything he can for a mechanical advantage. This includes winches, block and tackle, chain falls, come-alongs,

BASIC SKILLS REQUIRED

jacks and levers, skids and rollers, wherever and whenever possible. If a sizable vessel is really hard aground it may be impossible to tow it off the beach even with a powerful boat. But, by placing adequate anchors or deadmen offshore and rigging purchases (i.e., block and fall or chain fall, etc.), a sufficient strain can usually be brought to bear to take it off—sometimes in pieces.

5. *Patching.* Seamanship and rigging are skills, but patching is an art. You don't have to be a master carpenter to make sound plywood patches, nor a mason to fabricate a cement box over a hole in the bottom of a boat. But careful patching can save a lot of unnecessary effort. It must be done well enough to stop the leak and enduring enough to last until permanent repairs can be made, sometimes months later. The salvor has to know a little bit about making "hard" and "soft" patches, using underwater epoxy cement, hydraulic cement and concrete. Here again familiarity with hand tools is important as well as the ability to use them in the water and out.

6. *Diving.* Diving is not necessarily required on all salvages, but it is obvious that in many occasions work will have to be done under water. If you are a good boatman but not a diver and are seriously interested in engaging in salvage for commercial purposes, you'll either have to learn to dive or associate yourself with a good underwater man. If you are a pleasure boatman doing extensive cruising, you should at least be able to do a certain amount of skindiving. You may have to clear a line from a wheel or stop a leak, and some types of leaks can only be stopped from the outside.

7. *Mechanical Abilities.* The salvor doesn't have to be a first-rate diesel or gas engine mechanic, but should definitely have some mechanical skills. He must be able to look after his pumps and compressors and keep them running under ordinary conditions. He must be able to use hand tools and know what they are called since he will be using them. He will sooner or later be using clamps and fittings for water and air hoses, making patches where fastenings (nuts and bolts, nails and screws) are used, and utilizing wrenches, pliers, screwdrivers, hacksaws, pipe-threading dies and carpenter's tools. There is heavy emphasis on this subject in commercial diving schools since most professional divers are really highly skilled underwater mechanics.

One more thing: the would-be salvor should be in good physical condition. Salvage can be grueling work, and on occasion requires fairly strenuous effort. If you are not in good shape, it is advisable to do some swimming or running to save the old heart and lungs from some nasty surprises. Many active salvors may drink too much beer on occasion or smoke too much, but if they have been engaged in this type of work for a while, they are usually in good physical condition. If you get into salvage work and stick to it, you probably will be, too. This is one of the fringe benefits.

CHAPTER 3

The Gear

I have often been asked how much monetary investment a salvor might be required to have in equipment. The answer is, "not much." I have gone on salvages where the only thing I brought along was a change of clothes and a toothbrush. This is, of course, a qualified answer. You are not going to salvage anything with just a toothbrush.

Frequently a lot of equipment is required to carry out a successful salvage. The salvor does not necessarily have to own it, but if this be the case, he must know where he can either borrow or rent the equipment and material needed to do a job. Most salvors eventually accumulate equipment as they go along; nearly all tend to hang on to all sorts of miscellaneous gear left over from one job or another on the theory that it will come in handy sooner or later. They are probably right, but I call it the "Pack Rat Syndrome."

Realistically, the salvor should have as much equipment as he can afford. This would include a boat or access to one between 17 to 35 feet long. It should be sturdy and tough enough to engage in this work without the salvor being distracted by the occasional superficial damage that is bound to occur. The boat should be rigged so that towing is convenient and should carry an adequate towline, have a searchlight or hand-held spotlight, a good compass and some form of radio communication. It should also be fully equipped with the normal appurtenances and equipment one would expect to find on any well equipped boat.

If the salvor is a diver as well and diving is part of a particular job, he would normally have his SCUBA gear. If this is not the case, there should be a diving mask and swim fins aboard, and a wet suit if the water is cold.

Now let us consider the tools. You should have a basic set of hand tools. Carpenter's tools are needed to make patches and fasten them in place; wrenches for loosening and tightening nuts and bolts. You may not want to carry an expensive set of socket wrenches along, but there should be a good assortment of wrenches. Probably everything needed will fit into one fair-sized toolbox except for a saw and pry bar. This is a list of what is basically required:

Carpenter's handsaw (crosscut)	Asstd. open-end and box wrenches in larger sizes	Hand drill and bits
Hatchet	14" Adjustable wrench	Wire brush
Claw hammer	14" Pipe wrench	Scraper
Ball peen hammer	1 Set vise pliers	Wide putty knife
5/8" Cold chisel	1 Set water pump pliers	1 Sharp heavy-duty pocket or sheath knife
5/8" Wood chisel	1 Set regular pliers	1 Good flashlight
Asstd. screwdrivers, 1 heavy duty	1 Set side-cutter pliers	1 Hacksaw with spare blades
	Brace and bits	Crowbar

THE GEAR

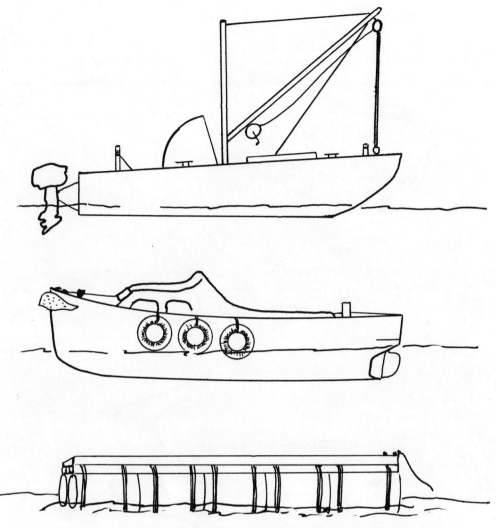

Fig. 1. (Top) Small steel or plywood work scow rigged for light salvage and towing. (Center) Twenty-six ft. workboat rigged for salvage. (Bottom) Work raft constructed of 4 x 4 timbers and 55-gal. drums. Raft has a plywood deck.

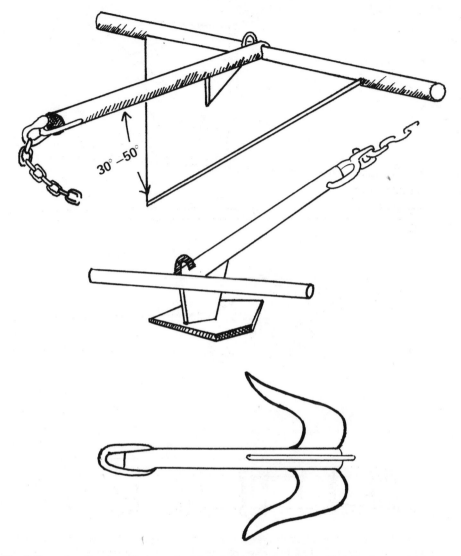

Fig. 2. Examples of anchors that may be fabricated from scrap metal. (Top) Sand or mud hook 30°-35° for sand; 45°-50° for mud bottom. (Center) Sand and grass bottom anchor. (Bottom) Grapple anchor for coral and rock bottom.

THE GEAR

This is a minimum assortment and will cost about $150.00. You can consider it your first-aid kit.

I suggest either borrowing or renting the pumps and compressors that may be needed, at least initially. When a salvor is well established, it will be economically sound to invest in this equipment. But if salvage work is engaged in sporadically, it is better to save the money. Salvage business is very often a feast or famine type of operation, especially for the beginner. Chain falls, "come-alongs," heavy-duty jacks, etc., can usually be rented or borrowed as needed.

Sometimes, especially if the salvor's boat is small, it is handy to construct a work raft or work scow. Then the gear doesn't clutter up the boat. The work raft can be dismantled and carried in the back of a pickup truck and reassembled on the beach. The workboat can tow it about as needed. If the raft is constructed of oil drums, each drum will support about 400 lbs. Five drums will support a ton, and a raft made up of ten drums is usually of sufficient size to carry the gear on most light salvage operations. The drawing shows how to construct one. (Fig. 1.)

Quite often anchors and grapples are needed in this line of work. It is a big saving if an arrangement can be made with a welding shop or machine shop to fabricate these from scrap metal. (Fig. 2.) For the active salvor, it may be worthwhile to buy a small AC welding machine and a set of gauges plus a cutting torch and learn to use them to fabricate your own anchors and metal fittings that are useful in this kind of work.

Keep an eye open for a cheap source of material such as plywood, timbers, cement, hydraulic cement, gasket material, used but usable cables, cable clamps and serviceable chain, truck inner tubes, empty oil drums, lengths of pipe, suction hoses, etc.

Used lumber yards, junkyards, tire repair shops and surplus equipment outlets are good places to look for bargains in these supplies. Material for salvage work is usually expendable, so it pays to know where to replenish it economically.

Chapter 4

Common Salvage Situations

When one considers the multitude of calamities that can befall marine equipment, it always comes as a surprise to find that there are really only about five different salvage situations that commonly occur (exclusive of fire). I will list them in order from the most common to the least.

1. *Simple Grounding*. This refers to a vessel aground or stranded and unable to free itself. The term "simple" indicates that although it may have incurred some bottom damage, the hull has maintained its watertight integrity and the vessel can be refloated without extensive repairs. The adjective "simple" can be misleading (similar to a "simple" fracture of the neck vs. a "compound" fracture of the "pinky finger").

2. *Aground and Sunk*. In most cases, vessels have suffered damage from driving onto a reef or bar, and are likely to be in poor condition. But sometimes boats are deliberately beached to avoid having them sink in deep water. They may not be damaged beyond the original leak that prompted the decision to beach them. In any event, it is obvious that the hull will usually require repairs before any effort is made to refloat it. I have observed cases where it was assumed that a vessel was undamaged (simple grounding), but after refloating, the vessel began to sink and was hastily shoved aground again, no doubt doubling the amount of damage to the hull. If the situation permits, it is wise to make an underwater inspection of the hull and check inside as well for leaks before attempting to pull it out into deeper water and perhaps suffer the embarrassment of having it sink there.

3. *Sunken Vessels*. This descriptive term covers a wide range of conditions. One vessel sunk in 100' of water may be unrecoverable. The same vessel, sunk in shallow water with decks awash, may be refloated (dewatered is the term) in 30 minutes. Vessels sink for various reasons. They may have struck a floating object (logs, heavy timbers) or may have broken a pipe leading to a thru-hull fitting, or suffered damage from a collision. It may also have been sunk alongside the dock by overenthusiastic fire fighters trying to douse a fire. It may require patching before refloating or raising, and at times it is best to raise the vessel first and make repairs later. This is usually the case in deep water. The vessel will have to be raised with flotation anyway, so the hull damage is immaterial at that time and can be repaired later in shallow water.

4. *Vessel Adrift*. A vessel might go adrift for a number of reasons. A dismasted sailing yacht may have exhausted its limited fuel supply; a motor yacht can suffer a power or steering gear failure; a trawler may break a shaft, or a vessel that has parted its mooring may set out to sea by itself.

COMMON SALVAGE SITUATIONS

Regardless of the cause of its distress or the nature of the remedy required, before anything can be done to help the vessel, *it must be found*. Sometimes this is no problem and at other times it might require an aerial search to locate it and (shades of the Bermuda Triangle) some never turn up at all!

Once the vessel is found, steps can be taken to either repair the damage, supply fuel, or tow it to a safe port, but while all these possibilities must be anticipated, nothing can be accomplished until the vessel is located.

5. *Submerged and Lost Objects.* This may be a case like the cruise ship that lost its anchor and chain mentioned in the first chapter, or it may be a sling load of prefab house components that falls overside from a freighter. It could be the brand-new outboard motor that just popped overboard from the transom of a runabout, or an old-fashioned anchor from a windjammer lost a couple of centuries ago (a real collector's item). Here again, the principle applies: Find it first, then decide how to raise it. In many instances you can mark the spot with a buoy while you leave to gather the necessary gear to recover the object. Sometimes it is better to take marks because if you buoy the object, it may be gone when you return. Taking marks will be explained in Chapter 9.

These, then, are the most common salvage situations. There are others no doubt, but if the salvor is prepared to cope with these, he has done just about everything he can to be ready and should be successful at this enterprise.

CHAPTER 5

Refloating Grounded Vessels

There is no such thing as a typical grounding situation as evidenced by the following description of some possible examples. One cruiser has an engine failure and drifts ashore onto a dangerous steep-to reef with a heavy sea breaking on it. The vessel will be smashed to kindling if not taken off promptly.

An unexpected gale causes a fine cruising ketch to part its mooring. It drifts ashore during a storm tide and grounds unharmed in a salt marsh a half-mile from the nearest navigable channel.

A great party is in progress aboard a large, deep-draft houseboat. The captain is also celebrating and takes an intracoastal day marker on the wrong side. The houseboat stops—so does the party.

A surplus landing craft (LCU) is being delivered from Charleston to Miami via the intracoastal waterway. The vessel has just cleared a bridge and the captain has to excuse himself. The helmsman is sleepy and so the vessel drifts out of the channel and goes aground. "No big thing"—it is getting dark, there is no tide, and the landing craft is built to go aground without damage. A decision is made to put out an anchor light and everyone can turn in for awhile. When daylight breaks, the sea gulls are walking all around the vessel. There is no lunar tide effect in this part of the Indian River, but there is a wind tide, and a fresh norther which promises to last a week is blowing. The water level will remain low until it blows out.

The question in each one of these cases is the same: How do you refloat the vessel? The answer in each case is different.

In the first situation, someone with a fairly powerful boat should get a towline on the grounded vessel and pull it off the beach. In this case it was tried and was a failure because the operator overlooked a few basic precautions. He should have had his own boat securely anchored before passing a towline to the vessel in distress. His boat did not have a tow bitt and was uncontrollable with the towline (which I suspect was his anchor line) made fast on a cleat on the after quarter. The line jammed on the cleat and could not be released. No knife was handy to cut the line and in the confusion, the towline fouled his propeller. Both boats were total losses. *If* he had anchored his boat securely, and *if* he had had a towline in addition to his anchor line, he probably would have been successful in freeing the other vessel. In any event, he would not have lost his own boat as a result of mistaken Samaritan principle.

The second case has a happier ending. A young man bought the boat from the insuror very cheaply. They had analyzed their own costs to recover it, paid off the owner's claim and then sold it to the highest bidder "as is, where is." There was only one bidder. The new owner promptly stripped it of all items that might be stolen: binoculars, wall clock, barometer, sails, pump, running rigging, etc. Then, since the vessel was lying on its side on a quite firm mud bottom, he removed most of the inside ballast, but shifted enough to keep the vessel heeled over. Three oil drums with the tops removed were rigged in line on a chain for a deadman or mooring and were jetted onto the bottom with a gas pump. Some long lengths of serviceable used elevator cable were acquired for scrap price and our salvor proceeded to drag the boat on its side back to the water's edge with a chain fall. He occasionally watered down the mud in front of the boat to keep it nice and slippery as he dragged it back to sea. It was difficult, time-consuming work, but he did the whole job almost unassisted and wound up being sole owner of a fine seagoing ketch for less than the cost of a secondhand outboard runabout.

A passing tug remedied the plight of the houseboat. The owner was appropriately grateful and the tug's captain and deckhand counted up their booty as they steamed away: about $100.00 in cash, two bottles of bourbon, a half bottle of scotch, two chilled bottles of French champagne and a tray of soggy hors d'oeuvres.

I was skipper of the LCU and we resolved the problem of the landing craft by jetting under it with the water pump and spotting the stern anchor (a 1500# Danforth) in deep water by lashing it to enough oil drums to float it, and towing it out into the channel with an outboard skiff and cutting the lashings. The big stern anchor winch pulled the vessel off with surprising ease despite the fact that a small tug (the same one that refloated the houseboat) had tried and failed to move it.

How does one refloat a vessel? In each case we have examined so far, refloating was accomplished by overcoming what professional salvors call "ground effect" and this varies according to the nature of bottom where the vessel is aground. (Fig. 3.) It is generally conceded in heavy commercial salvages that this factor may vary anywhere between 30% to over 100% of the actual weight of the vessel resting on the bottom. In the case of light salvage an experienced hand can often estimate this very closely. It sometimes helps if one remembers that 35 cubic feet of seawater weighs a ton, and estimate or calculate the number of cubic feet difference between the vessel's real waterline and where the water level comes on the grounded vessel.

Just as in regular commercial salvages, most efforts involve trying to tow the vessel into deeper water first. This is assuming there is no hull damage or that such damage has been repaired. If efforts to tow the vessel into deep water fail and there are no other vessels to assist in this fashion, then the salvor must rig up beach gear. Beach gear is a salvage term. It refers to the anchors, cables, and heavy tackles that are used to pull a stranded ship off the beach. These are capable of exerting strains of 30 and 40 tons or more,

Fig. 3. Some typical refloating operations.

and when overdone, can actually pull a ship apart. The light salvor can apply the same principle, but has quite a few more options than the commercial salvor. He can easily jet under the smaller vessel, use jacks to straighten it up, and shove skids or rollers under the keel to make it easier to slide. In big ship salvage it is assumed that the vessel will best come off in the direction opposite to the one in which it was going when it went on. In light salvage it is sometimes easier to take the vessel right across the reef, indeed it may be the only safe way if there is a constant heavy surge on the seaward side of

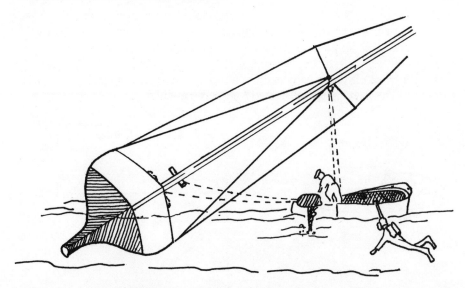

Fig. 4. A sailboat "heaved down" and towed to deeper water.

the reef and there happens to be (as often there is) a quiet lagoon to leeward of it. One sailing yacht was salvaged this way. The salvor took a boat cradle and sheathed the runners with sheet metal to make it skid easily over the rock and sand. He reassembled it under the boat and then skidded it about 150 yards right across the reef with beach gear that consisted of a big anchor made of scrap iron, some cable and a hand winch. Another sailing vessel sailed completely over a reef but sank in the shallow water inside as the portholes on the leeside were open. The salvor pumped the boat dry after sealing the ports. He kept the boat heeled down with a tackle on the mast and towed it into deeper water on its side. Once there he simply slacked off on the tackle to the mast and then towed the vessel in normal fashion. (Fig. 4.) Surprisingly enough, this boat suffered only a small amount of damage to its gel coat.

A problem encountered when using beach gear is the lack of a suitable place to make fast to on some vessels, especially motor yachts. Quite often even the anchor bitt or winch is only bolted or screwed to the deck and in such a fashion that a heavy strain will rip it off. The cleats back aft are not often sufficiently securely fastened to withstand this kind of force. When this is the case, it is usually best to strap the hull. A heavy line is passed completely around the vessel (more or less parallel to the waterline) and padded to avoid chafe at the stem and around the corners of the transom. This is usually held in place by lashings or by cleats nailed to the boat if it has a wooden hull.

It is often more effective to rig a similar strap around a vessel in order to tow it, unless its anchor bitt is well secured.

The salvor can now place his ground tackle. If there is mud bottom, a few empty oil drums with the top cut out can be chained together lengthwise or side by side and jetted in with a pump. (Fig. 5.) This will also work fairly well in soft sand. A mudhook welded from scrap steel can also be used. If there are good off-lying patches of rock or coral, an oversized grapple may work best. This, too, can be fabricated from scrap metal.

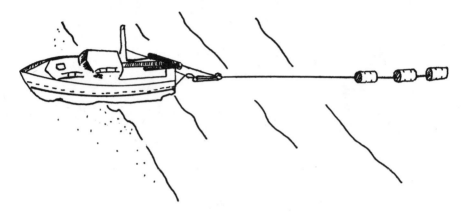

Fig. 5. Vessel being refloated with beach gear on mud bottom; 55-gal. drums are jetted in for a deadman.

It is best to use cable from the anchor to the block and tackle, chain fall come-along or winch that does the pulling. This will often be located on the vessel. The reason for this is that wire cable stretches very little and usable secondhand wire is often available from elevator companies, crane services, construction companies, etc., for the price of scrap. The next best choice would be either dacron fiber line or manila. Nylon stretches too much for this work, but it is great for towing.

The rigging methods are illustrated in Chapter 7. Now, assuming the vessel is hard aground on sand or mud bottom and the gear is rigged, get a set of tide tables published by the hydrographic office and concentrate your efforts at high tide. Sometimes a swell or wave action will help as it will lift the vessel and help break the suction of the bottom. If there is no swell, make one. Get a friend to run his powerboat as close as he can to the vessel to throw a good wake. (I know there are plenty of these powerboats around.) Any removable weight should be taken off if the vessel is stubborn about coming off. The remaining fuel and water, batteries, anchor chain and any ballast should be removed. The ballast alone might add up to several thousand pounds and 30% to 100% of that is not to be ignored. Jetting under the hull with a gasoline pump can remove a lot of material and will help break the suction of the bottom. You might also dig a trench to deeper water. In salvage work, just as in Love or War, all is fair and anything that works is good practice if it doesn't damage the equipment being salvaged.

CHAPTER 6

Dewatering the Sunken or Sinking Vessel

A vessel begins to sink whenever water enters the hull faster than its pumps can remove it. The remedy is obvious; simply reverse the process. The salvage term for this is "dewatering." This seems like a complicated way of saying "pump out" a vessel. But there is a little more to it than that and in some instances a vessel may not be pumped out at all. Air may be injected into the hull or some of the compartments and will displace the water.

For the moment, however, let us consider what is involved in dewatering without using compressed air. In order to accomplish this purpose, the pumping capacity must be increased and/or the leak should be reduced. The nature of the leak, the availability of additional pumps and the amount and kind of material on hand to effect repairs are all factors that will establish the salvor's priorities.

A vessel may sink from a minor leak that goes untended over a long weekend. A collision with a submerged piling may cause a leak that will sink a vessel in a few minutes if action to save it is not taken immediately. An open boat or one with a large cockpit that is not self-bailing can founder in seconds from heavy breaking seas and all the pumps in the world would not be able to save it.

The average light salvor can only prepare for these contingencies in a general way. He should carry as many pumps as can be used effectively (generally two 2″ gasoline-driven centrifugal pumps are sufficient). He must also keep enough material on hand to make fast effective repairs on steel, aluminum, fiberglass or wood hulls.

The normal procedure once the salvor arrives on the scene is to get the pumps going first and then start repairing the leak. But, if the leak is small and easily controlled with the pumps, repairs may sometimes be deferred until the vessel is towed or run under its own power to a location where it would be more convenient to effect the repairs.

If the leak is a large one, the salvor must either reduce it enough so the pumps can control it, or try to beach the vessel to avoid having it sink in deep water.

A matter of equal and sometimes greater importance than the pumping is the patching. This may be anything from a crude effort done in desperation to keep a vessel afloat in an emergency to a more durable repair of a semipermanent nature intended to serve until an occasion permits permanent repairs to be made at a later date (this could be several months).

The vessel's situation, as well as the size of the leak, will determine the kind of patch to be used. Large leaks stemming from a collision or something similar, are often checked by jamming a cushion, pillow, rags, blanket or anything that will fit into the hole, and then wedging it into place. Smaller leaks can often be stopped with wooden plugs, hydraulic cements like Water Plug®,* or by using plywood patches held in place by toggle bolts.

Most emergency patches are placed from inside the boat if the damage occurs in an area that is accessible. Occasionally a collision mat type of patch is used, or rags or plugs may be stuffed into an exhaust outlet or some other thru-hull fitting from outside when the pipe leading to it has carried away, but these situations are exceptional.

The semipermanent patches are ordinarily placed from outside except those for cement boxes. This usually requires diving and, for this reason, may not be possible in the open sea, especially if conditions are unfavorable. (Fig. 6.)

Much of the material used for patching can be applied interchangeably either for emergency or semipermanent patches; for example, emergency repairs can consist of:

1. Wooden plugs for small leaks and broken piping.
2. Water Plug®, a hydraulic cement which sets under water and against pressure, for small leaks.
3. Plywood patches with a gasket of foam rubber used with toggle bolts from inside.
4. Cushions, pillows, rags, etc., pressed against a large leak.
5. Waterproof tape and hose clamps for repairing broken pipes and hoses.

Semipermanent repairs, while still intended to be temporary in nature, may in some instances continue to serve for years.

In addition to the wood plugs and Water Plug® cement on wooden hulls, salvors can apply hard or soft patches. The distinction is sometimes hard to make, but a soft patch is generally considered to be a heavy canvas patch thoroughly coated with a bedding compound, old dregs of oil paint that have thickened, or asbestos roofing compound. This is nailed with flathead galvanized roofing nails or heavy copper tacks over the opening. Sometimes it is reinforced with strips of wood. If a piece of plywood is nailed down on top of the canvas it becomes a hard patch.

Sheet copper and sheet lead are used in the same fashion. Copper tacks can be used to fasten both copper and lead patches in place, but the galvanized roofing nails should only be used in conjunction with the sheet lead to avoid electrolysis. After the sheet copper or lead patches are in place, tapping the edges with a ball peen hammer will bend them into the wood and make a very tight seal. Sheet copper makes the better patch, but lead is easier to work with. Incidentally, plenty of older wooden boats owe an extended life to these types of patches as they will seal hard-to-caulk seams like the garboard turn of the bilge and open butts.

---

* Quick-Set Water Plug® is a registered trademark of Standard Dry Wall Products, 7800 N.W. 38th Street, Miami, Fl.

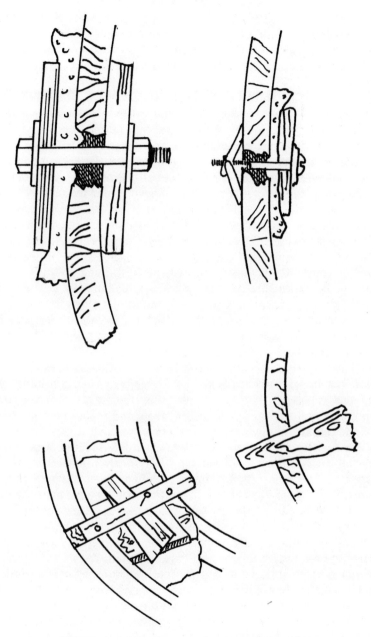

Fig. 6. Examples of emergency and semipermanent patches.

Plywood sandwich patches make strong effective repairs to holes and can close quite sizable openings. A sponge rubber or latex foam gasket works best, being placed between the outer piece of plywood and the hull of the boat. The two pieces are then drawn tight with a thru-bolt as shown in Fig. 6 (top left).

The single plywood patch can be used with a toggle bolt either inside or outside the hull. The best gasket here is the sponge type. It will always leak a little but generally makes a much better seal than any other material.

Cement boxes make acceptable repairs and are generally formed up from pieces of wood around a hole that has been temporarily stopped or partially stopped, then pure cement or concrete mix is poured up to the level of the top of the form. If there is a fairly strong flow of water from the leak, a copper tube is sometimes inserted into the mix after the leak has formed a channel. The water can drain off through the tubing which is then doubled over after the cement has set. Cement can be poured under water. If this has to be done, it is best to leave it in the bag and then open the bag in the box and pull out as much paper as possible without disturbing the mix. Plywood patches can be nailed directly onto wooden hulls, but they can also be attached to metal hulls with a stud gun of the ramjet type. Plywood patches can also be attached to fiberglass hulls by bedding them in underwater epoxy glue, and they can be held in place by a few self-tapping screws if pilot holds are drilled in the fiberglass beforehand. (Fig 7.)

Sometimes, trying to locate leaks can be a frustrating business. This is particularly true in older wooden boats. Leaks in fiberglass and metal hulls are usually pretty obvious; if they are not, then the leak probably originates around a thru-hull fitting. The propeller struts and rudder post are common sources and bobstay chain plates are another likely source.

Wooden boats tend to leak at the garboard seam, the bilge seam or chine, and at the butts. Centerboard wells are vulnerable to worm damage, as is the rudder post. Stopwaters often rot out after a few years and provide a lot of mysterious leaks that are never found. These are located at the scarfs in the rabbet line in keels, stems, deadwood and horn timbers. They can usually be drilled out and replaced without removing the plank, but for temporary repairs, copper sheathing across the scarf and a cement box are the best ways of stopping it.

The salvor who anticipates getting much dewatering work must have adequate pumping power and tools and material to make patches. This is just as well, for unfortunately many of the salvor's own vessels are the marine equivalent of the "mechanic's special," and remain afloat as a result of constant if not careful attention.

However, the boatman who intends to venture offshore could do worse than emulate the salvor, to the extent of at least carrying additional pumps to those normally found aboard a pleasure boat and having the material on hand to make fast secure repairs to whatever leaks may develop.

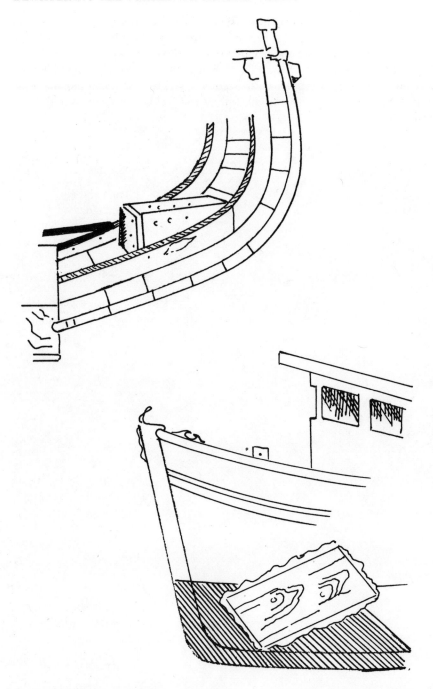

Fig. 7. (Top) Cement box. Repair to damage in the bilge of a launch. (Bottom) Plywood and canvas patch over a large hole in trawler's bow.

The hull of the offshore cruiser should also be accessible from inside (many are not) so that leaks can be repaired without having to go over the side at night or in rough and possibly shark-infested waters.

I've witnessed a few maritime "post mortems" and in most instances, if the vessel had been equipped with adequate pumps and patching material and if the interior had permitted full access below, disaster would have been averted.

CHAPTER 7

Raising Submerged Vessels

A vessel resting on the bottom, in any depth of water that covers it completely, can be raised by one of four basic methods:
1. Pumping.
2. Raising with flotation.
3. Raising with internal air.
4. Physically lifting.

Each method has its advantages and disadvantages, and quite often the circumstances make it impractical to consider more than one or two of these methods.

It usually comes as a surprise to the average boatman that a vessel completely submerged can be pumped out. But, in shallow depths, underwater pumping may be the easiest way to raise a boat. The motor yacht, unless it is exceptional, does not lend itself well to this practice for two reasons—it usually has too much glass in the wheelhouse and the cockpit deck will probably have leaks galore. Sailboats are often best raised this way. Most sailing craft of any size are reasonably watertight above deck, with small self-bailing cockpits and no wheelhouse glass to break. (Fig. 8.)

The limitation in this technique is the depth, since in order to pump out a boat, a source of air from the surface must be supplied so the water can be removed. The air is at surface pressure and the hull of the vessel is subjected to the pressure of the surrounding water and can be crushed if there is too great a differential between the two. If, for example, the boat lies in 35' of water, the external pressure would mount to about 15 p.s.i. once pumping started, and on the conventional pleasure boat hull, this would probably ruin it. Commercial vessels built of steel, such as small tugs, barges, etc., could probably withstand the pressure, although the portlights might have to be covered. For yachts, especially fiberglass and wood, 7' to 8' would more than likely be as deep as you might wish to use this method.

The mechanics involved are straightforward: patch the leaks, provide a source of surface air and a watertight entrance for the suction hose—then start pumping.

If pumping is not feasible, the salvor must use compressed air in one form or another, unless he has a platform large enough to permit hoisting the wreck bodily from the bottom.

In the case of most small vessels, the salvor can use compressed air to fill lift bags, drums and inner tubes secured to the hull until he provides enough external flotation to overcome the negative buoyancy of the vessel.

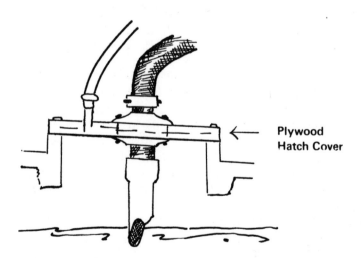

Fig. 8. (Top) Pumping a submerged vessel. (Bottom) Arrangement for passing suction and air supply hoses through a hatch.

RAISING SUBMERGED VESSELS

Many vessels have large enough fuel and water tanks to float them, or at least contribute quite a bit of the buoyancy required to raise them. These tanks are usually securely fastened in place and can be filled with air from the standard fitting. Fuel tanks usually have baffles and the fill pipe commonly extends almost to the bottom. The air hose can simply be inserted in the tank after the vent has been plugged, and the air pumped in will void the liquid contents. You may wish to pump in some Gamelen or other emulsifying agent in order to avoid an oil spill. This can be pumped in via the same air hose. Water tanks may not have baffles and may not have the extended fill pipe found in fuel tanks. A bushing can be fitted to the filling line to take the air hose, but the discharge line and drain plug should be removed in order to void water and the excess air as it expands when the boat starts to rise.

The source of the salvor's air will in most cases be a portable compressor driven by gasoline or diesel engine. It need not be very large; 35 cubic feet of air will lift one ton and a very modest unit will deliver this volume in a minute at 40 p.s.i. (nearly three atmospheres). Even small diaphragm compressors are adequate for small salvage jobs and having any sort of compressor on the job is cheaper than trying to fill your flotation gear from air bottles.

A small manifold to supply air to the lift gear can be fabricated from pipe fittings and valves available at any hardware or plumbing supply store. This is an important piece of equipment as it helps the salvor maintain the equilibrium of the vessel while it is being raised. This is usually accomplished by supplying air to both ends and on both sides of the boat, working from the forward and aft towards midship. (Figs. 9 to 11.)

Sunken barges and some metal hulls can often be raised more easily by injecting air into them. In order to do this, the decks must be tight to prevent the air from escaping. It may be immaterial whether the hull bottom leaks or not, as the air must have some place to escape as it expands. The 1,200-ton tank barge referred to in the first chapter had holes all along the chine but did not sink until the decks became porous. The salvor rigged up a small manifold to control the flow of air to the tanks and then systematically plugged all the holes in the deck. The barge floated on the bubbles in its cargo tanks.

Air can be injected from the top as well as the bottom by installing "standpipes" in the hatches. On some leaking barges permanent standpipes have been installed that can be used for pumping out the barge or injecting air into it. The principle of the standpipe is really very simple. As air is injected into a tank the air displaces the water which spouts out of the standpipe. Standpipes are usually installed deep in the tank, but at times it helps to calculate the actual volume of air required to float a vessel and then saw the standpipes off to the proper lengths, so that excess air will void from the tank after the correct amount of water has been displaced. Standpipes are

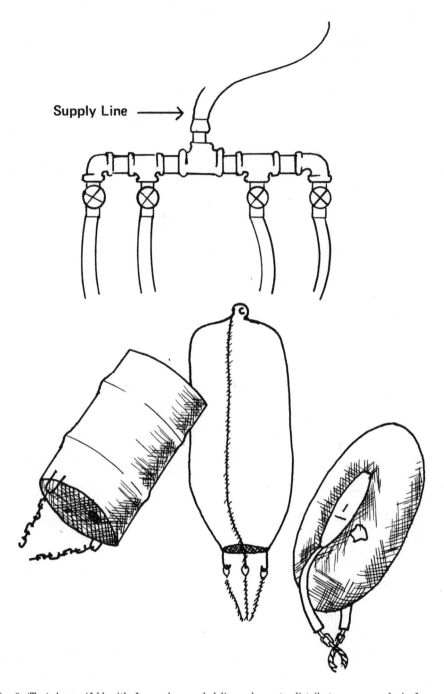

Fig. 9. (Top) A manifold with four valves and delivery hoses to distribute compressed air for flotation. (Bottom) Flotation gear: Lift bags, inner tubes and 55-gal. drum. Drum has "U" bolts welded top and bottom for securing it in place.

best made by fitting either a metal nipple or a coupling in a steel plate or heavy piece of plywood cut to the shape of the hatch to be covered, and fitted with a heavy neoprene gasket as an airtight seal is essential for this method to work right.

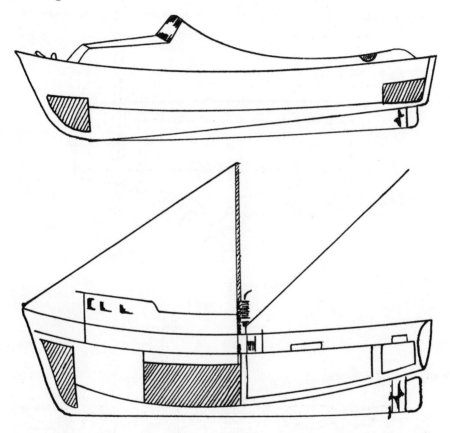

Fig. 10. Residual flotation in hulls. (Top) A typical motor launch with water tank forward and fuel tank aft. (Bottom) Trawler with fuel tanks located port and starboard midship and a water tank located forward.

After this is done, metal or heavy duty PVC pipe can be screwed onto the nipple or coupling and inserted into the hold, and the hatch secured. I usually inject the air by simply passing the air hose right down the standpipe which should be 2″ or 3″ in diameter. The flow of air to the compartments is then controlled by the manifold. It must be borne in mind that air expands rapidly when a vessel is rising off the bottom. As its volume is inversely proportionate to the pressure, the volume would double if the pressure were reduced by one-half. The salvor must also keep himself and his workboat clear when raising a vessel on air. It can come to the surface in a rush and carelessness will cause injury and damage. (Fig. 12.)

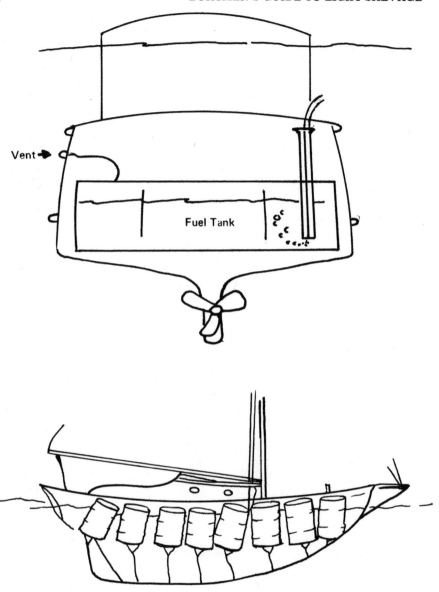

Fig. 11. Examples of external and internal flotation. (Top) Method of utilizing a fuel tank in the stern of a cruiser. (Bottom) A sailing yacht raised with external flotation.

RAISING SUBMERGED VESSELS

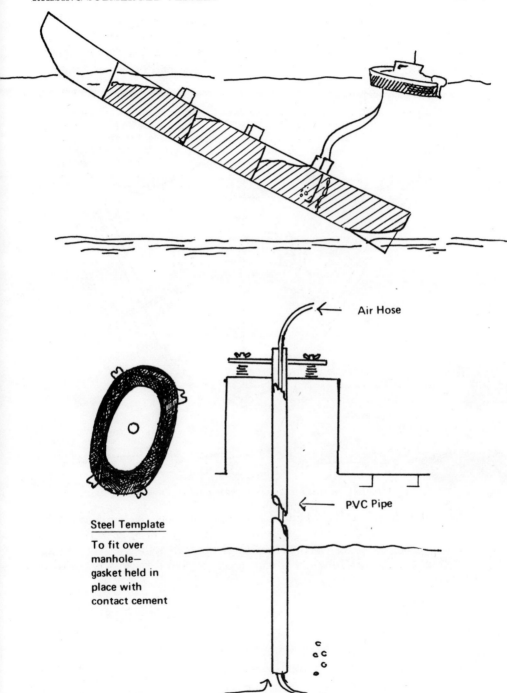

Fig. 12. (Top) A barge being raised by compressed air injected through the standpipes. (Bottom) Method of installing and utilizing standpipes.

A salvor may find occasions when it is practical to raise a vessel bodily. In order to do this he must have a stable platform capable of bearing the weight of the submerged vessel. A pair of lifeboats lashed together catamaran fashion, or a float made from dredge pipe, might suffice for vessels up to

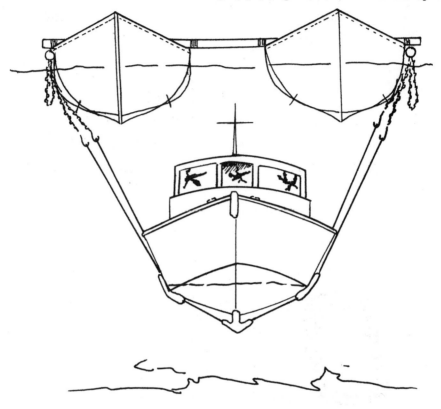

Fig. 13. Two lifeboats make a platform to raise a motor launch.

about 40′ in length. (Fig. 13.) Slings can be passed under the vessel and it can be lifted just as if it were in a shipyard. There would have to be chain falls or come-alongs at both ends on the slings to avoid rolling the boat. Some small boats may have lift rings strong enough to permit them to be raised by these.

CHAPTER 8

Vessel Adrift—Rescue Towing

When a vessel is adrift and requires a tow to reach a safe port it is called a rescue tow. This is also considered an act of salvage. But before a salvor can do anything to assist the vessel, he must find it. He may not have to go as far offshore to carry out his operation as a large salvage tug, but the basic problems are pretty much the same. In order to succeed at this undertaking the salvor must have the following information:

1. A comprehensive description of the vessel and its name.
2. Its most recent location.
3. The nature of the damage or reason for the vessel being unable to carry on under its own power.
4. The kind of communication equipment it has and whether this equipment is working.
5. Try to ascertain whether the vessel has a searchlight and day/night distress signals.

Needless to say, the salvor's vessel should be equipped for towing. Necessary gear includes:

One good nylon towline (in addition to an anchorline).
One set of wire bridles and chafing gear.
One good heaving line.
One portable gasoline pump, with hoses.
One extra fully charged, heavy-duty 12-volt battery.
Spare fuel (gasoline and diesel in 5-gallon cans).
Plenty of spare patching material and a few expendable hand tools.

The salvor's boat should be well fendered and have a radio and searchlight. The compass should be corrected and there should be instruments for adequately plotting the distressed vessel's supposed position and his own track to reach it or search for it if this becomes necessary.

The first and most important step is locating the vessel. Sometimes this is easy, often it is difficult; but establishing radio contact directly with the vessel helps a lot. Secondhand information is often garbled and inaccurate.

If the vessel's location is not precisely known, radio bearings will help to locate it. At night a powerful searchlight can be directed upward and frequently can be seen for a considerable distance, especially if there are low-lying clouds to reflect the light. The day and night distress signals can be used if the salvage vessel is within reasonable range. Quite often the boats are anchored waiting for help, but if they are offshore and cannot anchor, the salvor should have knowledge of local tide and currents and take the wind into consideration.

If the salvor is obliged to conduct a search for the vessel he should get a little to weather of the probable area of its location and, depending upon the speed his vessel makes and his height of eye, lay out a ladder-type search pattern on his chart and adhere to it. (Fig. 14.) His height of eye is important because the distance to the horizon is equal to 1.14 times the square root of his height of eye. For example, if his eye level is about 9', his horizon lies at $1.1 \times \sqrt{9}$ or 3.3 miles. If the top hamper of the other boat is about the same he will be able to barely see it six miles off. Therefore, the legs on the search pattern should not be more than eight or ten miles apart. If the legs are ten miles apart and ten miles long, he will cover an area 20 miles wide and if his speed is ten knots, he will make five knots toward the location of the vessel he is seeking.

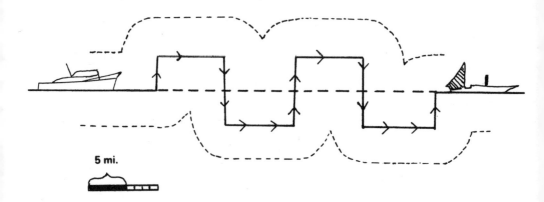

Fig. 14. Ladder search pattern. Broken line indicates probable line of drift of distressed vessel. Solid line indicates track of salvage vessel. Dotted line encloses the area searched.

It may be that the problem can be resolved without taking the vessel in tow. It may only need a battery to start the engines or a pump to dry the bilge—or it simply may have run out of fuel. Patching material may come in handy in case the vessel has a serious leak.

Any pleasure boatman contemplating going offshore should take a look at the list required of a salvor. It would cost little additional money to keep these stores on board, and there is an increasing likelihood these days that this equipment would eventually prove its worth. As the adage says, "Hope for the best, but prepare for the worst."

Incidentally, a fairly small low-powered boat can tow a much larger vessel if the operator uses the wind and current to advantage. In this case the closest safe port may not be the best choice, especially if it entails going to windward during heavy weather. It may take ten hours to gain 15 miles to windward, and only require three or four hours to cover 20 miles to leeward.

Chapter 9

Submerged and Lost

I didn't know it at the time, but my first successful salvage was a "submerged and lost." I was about 14 years of age and had been skindiving in a creek bordering a golf course. I found a golf ball, promptly sold it to a golfer, and spent the evening at the movies on my earnings. It was not until many years later that I appreciated the implications.

Many divers turn up things of value or interest in the course of their activities. Much of this is recovered. There may be artifacts as large as an old sailing ship anchor or as small as the antique bottles found in some areas. My own children, fledgling salvors at heart, dive for fossils (including sharks' teeth) in freshwater streams in central Florida. But the commercial salvor more often than not knows what he is looking for and more or less where it might be found.

This is really the crux of the problem—"more or less" is not good enough. The salvor must find the item (know exactly where it is) in order to recover it. There are various methods of searching for something that is lost under water. The two principal methods used are an underwater search and by dragging.

It is quite difficult for a diver to orient himself under water. There must be some reference point established so that he can make a thorough examination of an area, and then go on to another adjacent one secure in the knowledge that he is covering the ground thoroughly, but not repeating himself. If the water is clear enough and the object sought is big enough to be seen from the surface, a rather thorough search may be carried out from a boat with a glass bottom in a well or a water glass, but the best way is to tow a diver slowly astern of the boat, preferably on one of the underwater sleds used for this purpose. The diver's range of vision is much wider than either a water glass or glass-bottom boat. The boat operator must keep track of the boat's position. This can be done by buoying the area or using landmarks for reference points.

If the water is not clear enough for this, then the diver must swim along the bottom in ever-increasing circles around a drop line. (Fig. 15.) This is a buoy and anchor, and the diver holds by hand a length of light line (with knots tied at even intervals) attached above the anchor, and at every complete revolution around the drop line he will slack as much line as visibility permits. Once an area has been searched, a buoy of some kind (a large plastic capped bottle is fine) should be left on the spot and the drop line shifted to a new search area. The adjacent areas should overlap slightly to avoid the object being overlooked due to poor visibility. A very thorough search can be carried on this way, but it will be time-consuming.

If the bottom is smooth, with few obstructions, two boats working together dragging a wire on the bottom can cover a lot of area. There should be a sturdy towline attached to a weight and the wire is stretched across from one weight to the other. One boat should be the leader and make the course,

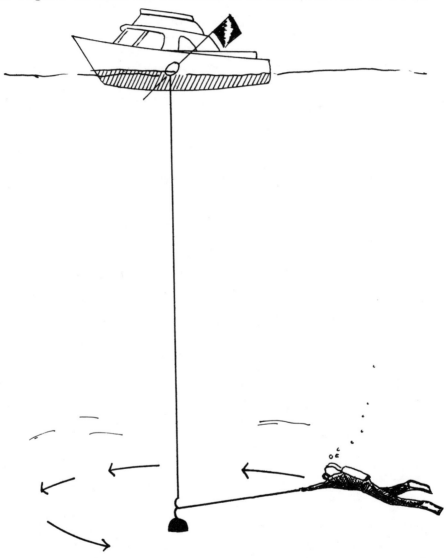

Fig. 15. Diver carrying on bottom search around a "drop line."

while the other maintains a parallel course some distance away. If they are careful they can search a tremendous area of bottom in a day. The area should be searched on reciprocal courses that slightly overlap.

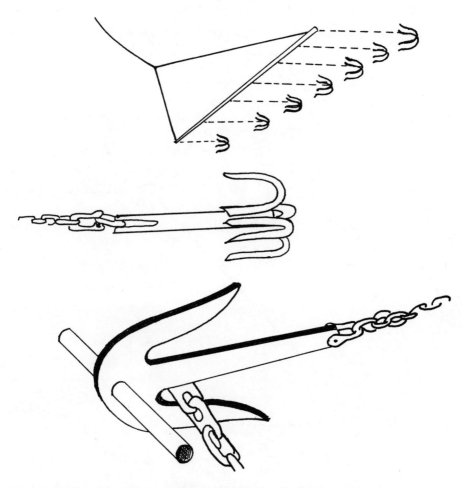

Fig. 16. Dragging devices for locating submerged objects. These may be fabricated from scrap steel. (Top) Grapple "rake" for small objects. (Center) Grapple. (Bottom) Anchor hawk for recovering anchors and chains.

Dragging with grapples is a useful method; either a single heavy grapple can be towed behind when one is looking for a cable or hawser that is lost, or a "rake" 10' to 15' wide towed on a bridle with a number of smaller grapples attached can be used for other smaller objects. This would include light chain, anchors, anchor lines, lost moorings, etc. When a heavy chain is lost and diving is unproductive, an anchor hawk can be made out of heavy steel plate as shown in Fig. 16. This will hook up a chain even if it is completely covered with mud or sand provided it hasn't sunk in too far.

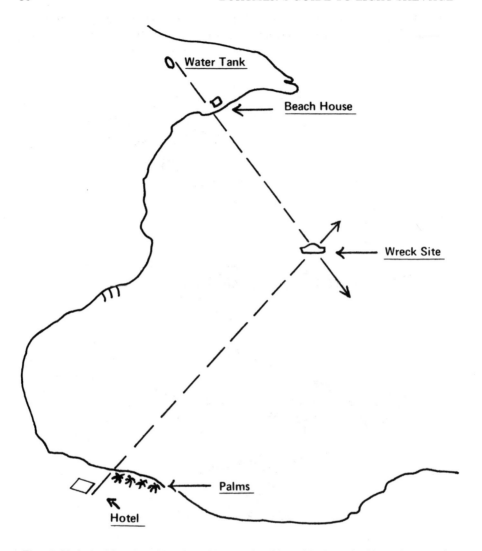

Fig. 17. Method of locating object by taking marks. Often this is preferable to buoying the object.

A hit-or-miss dragging operation may turn up the object sought, but it is far better to be painstaking and be certain of finding it by carefully running a grid of courses.

The method of recovery will depend entirely on the nature of the object once it has been found. A large object is usually raised with flotation and then towed in. If it is small enough it may be hoisted aboard the workboat. Often it is important to be able to establish the location of a submerged object without marking it with a buoy. Commercial fishermen do this all the time to avoid having other fishermen clean out their "fishing hole" or reef. (Incidentally, quite a few wrecks have been found to be "fishing holes" as the wrecks attract large concentrations of bottom fish.) Pinpoint locations can be established by the use of ranges much more accurately than by compass bearings. I have returned to locations within a few yards of an object up to 12 miles offshore! Ranges are just two conspicuous landmarks that fall in line. Usually two sets of ranges are used; one is called the running mark and the other is called the stopping mark. The boat simply steams down one range until the other range falls into line. I usually make a sketch of the marks on a piece of paper and also plot them on the chart if the range marks are shown. Range marks can include radio towers, water tanks, points of land, a lighthouse, clumps of trees, buildings—anything distinctive that is on a fixed location. (Fig. 17.) The one in front is lined up with a suitable mark behind it. This can be a mountain range or anything else that is convenient and readily recognized. The ranges should be as close to 90° apart as possible, to minimize slight errors that occur. Single ranges can be used effectively in conjunction with a recording depth sounder if the bottom has a noticeable configuration. They may also be used with vertical and horizontal sextant angles and a range can be crossed on LORAN lines for fixes far offshore.

Pat Boatright and I recovered an anchor that had been lost for two years. It came from a big barge that had been anchored between cable areas. In order to be certain that the barge was anchored clear of the cables, ranges had been taken and plotted on a chart. Other divers had searched the general area quite thoroughly and swore that the anchor was not there. Pat located it on his first dive and complained because he had to swim 35 feet to buoy it!

CHAPTER 10

Seamanship, Rigging and Beach Gear

Seamanship is a difficult term to define. It implies a proficiency at a variety of skills and the judgment to employ them effectively even in difficult or dangerous situations. Acquiring the skills is usually easy enough (they are basically of a mechanical nature), but judgmental wisdom is more likely the product of experience.

Marlinspike seamanship is the area where most non-professional boatmen's deficiencies are most apparent. This is especially true in salvage, as the salvor must be able to make a variety of knots, bends, hitches and splices and be capable of rigging a set of falls (block and tackle). The salvor also should know how to throw a heaving line, make secure lashings and slings, apply chafing gear, and pass a stopper on either wire or fiber rope.

The working lines that are normally used on a salvage should be either nylon, dacron, dacron blends or manila. The nylon is really fine for a towline or an anchor line for your boat, but it is not good to use in setting up purchases or lashings as it is too elastic for this purpose.

Dacron is a good line for just about every purpose. It is almost as strong as nylon and does not have as much stretch. For this reason, it is not as good a tow hawser but it is excellent for blocks and falls, slings, lashings, etc.

Splices in both nylon and Dacron require at least four tucks. I believe five tucks are better.

Dacron blends (Polydac, etc.) will suffice, but they are expensive and are about twice the strength of manila. However, they are lighter and are supposed to float.

Manila is only about 40% as strong as Dacron, but is much cheaper and in the sizes used in light salvage, can be utilized just about everywhere. A splice in manila line requires only three tucks.

Polypropylene and polyethylene are really too slippery in small sizes for use in salvage work. The knots frequently slip. Avoid using these if possible.

When line is sold in 600′ to 1,200′ lengths, small size line often comes on a wooden spool. Larger diameter line usually comes on a coil and is often covered with a burlap lashing. Big hawsers, 2″ diameter or bigger, normally come on large wooden spools. When a line is on a spool it can be removed either by rolling the spool back and forth so that the line is faked out or by allowing the spool to spin freely about an axis while the line is pulled off. If regular lay line comes in a coil, the line should be removed by reaching inside the coil and pulling up the bitter end (this is usually tagged) from the bottom inside the coil. If an attempt is made to uncoil regular lay line from the outside, it will be a mess.

Nearly all new regular lay line has a few kinks. The easiest way to get them out is by "thorough footing" the line. It should be coiled down left-handed (counterclockwise); the coil is then capsized (turned over) and the bitter end pulled up through the bottom and coiled down in the normal fashion (clockwise). The braided lines can be coiled by either method.

In the past, small sizes of line were called by thread sizes from 6-thread up to 21-thread, being from about 1/4" diameter to about 1/2" diameter; after that, they were called by the diameter up to about 2" diameter and then they were called by the circumference. This practice still applies.

Line is commonly measured in units of 6 feet, called fathoms. A seaman usually knows his arm span and with practice can measure off a fairly uniform fathom. It is good to know how to do this as it saves time trying to measure a long length of line.

Knots are for tying eyes in a line and joining lines. A seaman really doesn't need to know how to tie many knots, but he should know how to make them by feel so that at night or under water he can tie them securely. Bends are for bending lines together, often lines of different sizes. Most of the knots that a salvor should know are illustrated in Fig. 18.

Hitches are used for making lines fast, either around a pole or piling, and some of them are used for fastening to cargo hooks. A hitch around a post should be able to be cast off when under a strain. A clove hitch is seldom used as it will loosen under varying strain and may jam tightly under a heavy load. The best one for this is a round turn and two half hitches (several round turns can be taken if there is a heavy load that must be slacked). A towboatman's hitch is useful for securing a hawser on a capstan or around a single bitt; the blackwall hitch is for attaching to a cargo hook, as is the "cat's paw." (Figs. 18 and 19.)

Stoppers are short lengths of line or chain that are used to hold the strain on a line or cable under tension. This is called "stopping off" and is done so that the line or cable can be slacked and made fast on a bitt or cleat without paying out. This is also done on cables or anchor rodes on beach gear when the length of the tackle has to be readjusted. Carpenter's stoppers (or one I designed) made up of wire clamps and reinforcing rod (rebar) are used to attach the tackle of the beach gear to the anchor cable. (Fig. 20.) Rope stoppers are made from the rope, preferably manila, as it has rougher texture and holds well. The stoppers should be of smaller diameter than the line they are stopping off. Chain stoppers are used on cable, though a rope stopper will hold, but with only a limited amount of strain. (Fig. 21.)

Single stoppers, rope or chain, are usually a series of half hitches a distance apart so they will not jam when they are made up on the hauling part of the line being stopped. A rolling hitch is also used with rope stoppers. This is sort of a double half hitch and it will occasionally jam. Double stoppers are sometimes used. These are simply wrapped in opposite directions

(text continues on page 43)

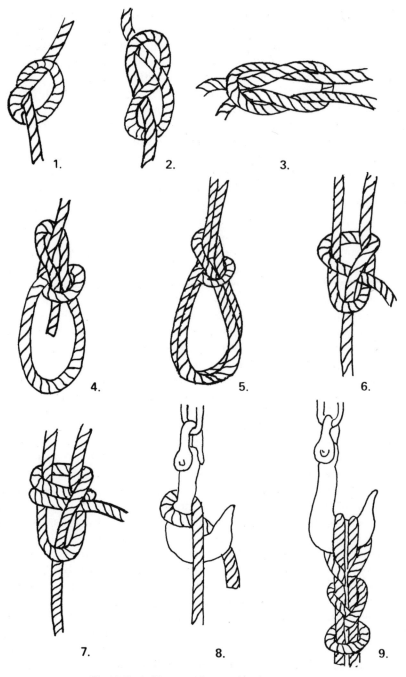

Fig. 18. Typical knots a salvor needs to know.

1. Overhand knot.
2. Figure eight knot.
3. Square knot.
4. Bowline.
5. Bowline on a bight.
6. Sheet bend.
7. Double sheet bend.
8. Blackwall hitch.
9. Cat's paw.

SEAMANSHIP, RIGGING AND BEACH GEAR 41

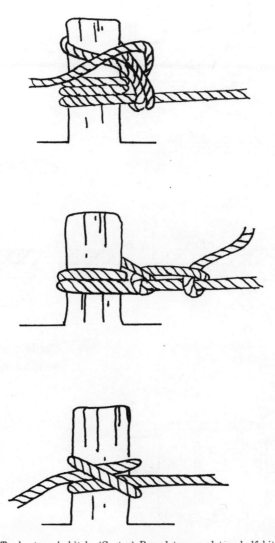

Fig. 19. (Top) Towboatman's hitch. (Center) Round turn and two half hitches. (Bottom) Clove hitch.

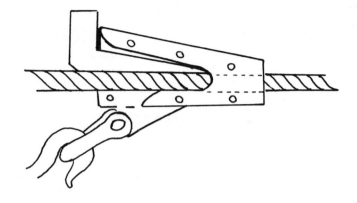
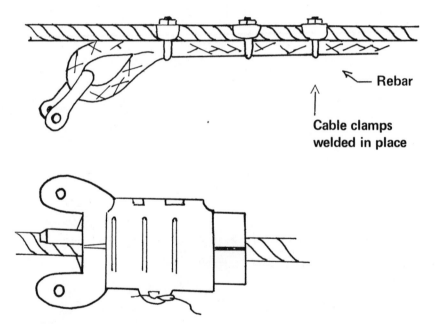

Fig. 20. Examples of "carpenter's" stoppers used to attach block and tackles and chain falls to cables. (Top) The stopper is assembled around the chain and a wedge is driven home to make it tight. (Center) The stopper is simply clamped to the cable and is shifted by loosening the cable clamps. (Bottom) A patented stopper used by commercial salvors. It opens for removal and locks up by driving in the wedge.

SEAMANSHIP, RIGGING AND BEACH GEAR

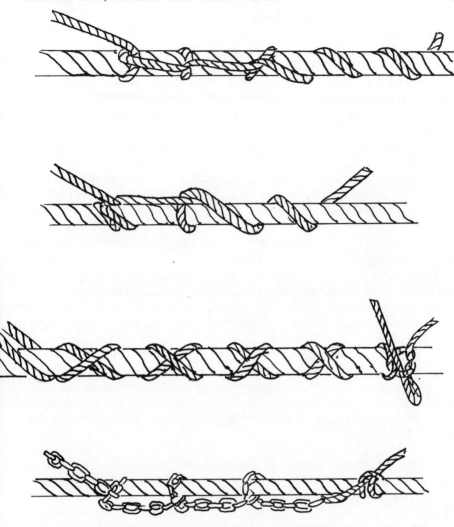

Fig. 21. Methods of stopping lines when under a strain. The chain stopper is for use on cable.

around the hawser or cable. On chain stoppers there should be a manila pigtail which can be tied off.

Straps, or strops, as they are sometimes called, are either short lengths of line short-spliced into a circle or with an eye spliced at either end. Slings are more or less the same thing but longer. Straps are used around lines and spars hanging on a cargo hook, shackle or sister hooks. Slings are larger, for passing around a barrel, a keg or a crate that one desires to lift or drag.

Cable straps and slings are used for heavy lifts. Some of the different types of slings and their uses are illustrated in Fig. 22.

Lashings are used to secure things to each other. When two pieces of line are lashed together, this is called "married" or "seized" together. Wire rope is usually seized whenever it is to be cut unless an acetylene torch is used to cut it, this prevents the end from unlaying. The ends of fiber rope can be wrapped with a good grade of tape, preferably waterproofed for the same purpose.

Most nonprofessionals cannot splice wire. But the Flemish or Dutch eye splice illustrated in Fig. 23 makes a neat strong eye that can be seized or secured with one wire rope clamp. This is handy for making straps and slings and turning an eye quickly in cable. Thimbles should be spliced into the eye of any line riding on a shackle. But in many instances it can be disposed with when using cable, as it is tougher than fiber line. Cable clamps can be used for making eyes in wire rope. The saddle of the clamp always goes on the standing part.

"Blocks and falls" or tackles are used to increase the pulling power. It works like a reduction gear. If, for example, a two-fold purchase is used, it means that two feet of line must be pulled in order to lift or drag an object one foot. If the object weighs 100 pounds it will require 50 pounds of pull plus the allowance for friction to do so. For salvors, a six-fold purchase is about all that is practical because overhauling the blocks (stretching the falls) requires effort too. However, by compounding the purchase called "a tackle on a tackle," the mechanical advantage is multiplied again, i.e., 6 x 6 = 36. Ten percent for friction is generally allowed for each sheave and is added to the gross weight of the object to be lifted, i.e., 1000# + 60% for friction = 1600 ÷ 6 = 267#.

Salvors will often compound a purchase with a winch, a chain fall or come-along. In other words, a ton and a half come-along on the hauling part of a six-to-one purchase with allowance for friction can pull a load of about six tons. If a three-ton chain fall is used, it would double this amount; this much pull would equal the amount of pull developed by a 1,000 hp tugboat!

Figure 24 shows a typical block and fall arrangement. The power or ratio is determined by the number of parts (lines) leading from the moving block. There is a 10% allowance for each sheave for friction. Surplus lifeboat block and falls work well and are usually reasonably priced compared to new unused blocks.

Fairleads are sometimes required in order to lead the hauling part of a tackle to a winch or anchor windlass. Single-sheave snatch blocks are handy for this.

(text continues on page 48)

SEAMANSHIP, RIGGING AND BEACH GEAR

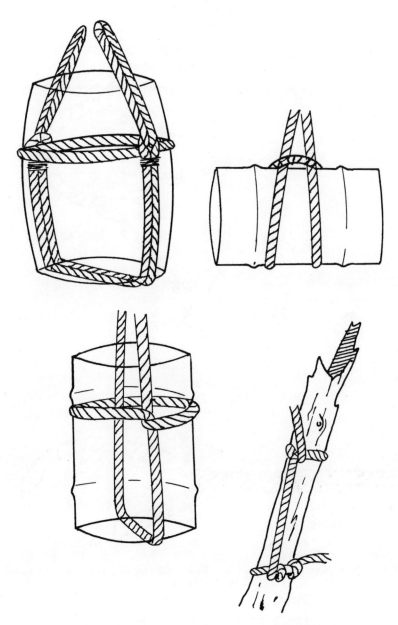

Fig. 22. Various types of slings (Top, l. to r.) Double barrel sling; side sling on a drum. (Bottom, l. to r.) Single barrel sling; timber hitch for towing or lifting a spar.

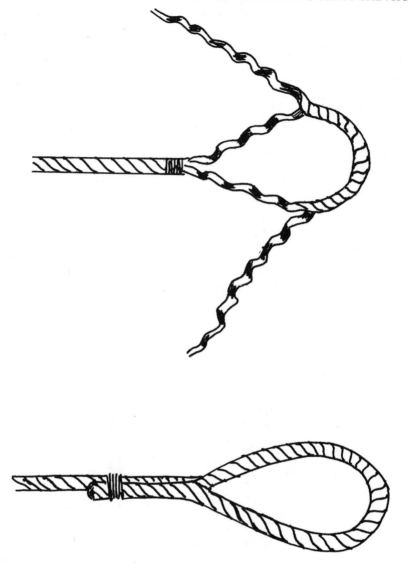

Fig. 23. Flemish eye splice or "Dutchman" in a cable. Ends may be seized as shown, or secured with cable clamps.

SEAMANSHIP, RIGGING AND BEACH GEAR

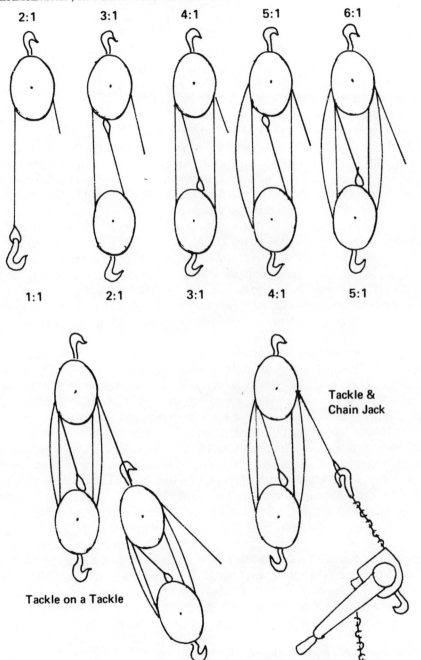

Fig. 24. Block and tackle power ratios are indicated at top and bottom of each fall. Compounding the power is shown at the bottom.

Lashings are required to secure drums and inner tubes for flotation, to lash spars together, or to form and sheet legs for a lift. Lashings can be used for repairing cracked or damaged mast or booms. If a vessel hull is strapped for towing or kedging off with beach gear, lashings will probably be required to hold it in place. There are several kinds of lashings illustrated, but the salvor must choose what is appropriate for his situation.

There are some quick-release methods that are handy to know. Use of toggle pins, pelican hooks and some reef knots will allow the salvor to let go or cast off a line in a hurry, even when it is under a heavy strain. This is often convenient. (Fig. 25.) Tapered hardwood bedlegs can be bought at a hardware store and make good toggle pins for this purpose. Old wooden fids can also be used. A hole can be drilled at the heavy end for a lanyard if desired. This allows the object to be tripped from a distance.

Someone once stated that if a single man had a lever long enough and a fulcrum, he could move the earth. The author of this observation was probably a rigger. To the uninitiated usually the word rigging is taken to mean a vessel's rigging when used as a noun and the repairing or replacing of it when used as a verb. But "rigging" is also the practice of moving heavy objects, and this is the context in which the word is used here. It is a shoreside word of nautical derivative, so it should not come as a great surprise that many riggers have a seafaring background.

Riggers use anything they can to reduce friction and provide a mechanical advantage. This includes skids, rollers, wedges, prybars, levers, and a variety of jacks. They also use chain falls, block and tackle, come-alongs, and winches. There is a broad area where rigging and seamanship overlap, but riggers employ a few techniques that are seldom required of seamen.

One salvage operation that I recall was all rigging work. A landing craft had been deposited by hurricane winds far above high water. The salvor jacked up the vessel and placed cribbing and skids under it with pipe rollers in between. There wasn't much of a grade to the beach but in spite of this his principal problem was restraining the vessel while working it sideways down the beach. He finally had a bulldozer launch the landing craft for him when it reached the water's edge. I don't think he even got his feet wet. (Fig. 26.)

If the bottom is rough, a couple of skids under a vessel's keel, especially if they are covered with sheet metal, may reduce the friction enough to make it an easy pull. But the salvor must weigh the advantages. At times the strut, propeller, shaft and rudder should be removed; a boat's bottom is usually pretty smooth, and will slide easily if the appendages are removed to prevent them from hanging up on rock or coral. Often a vessel is better off driving high and dry on reef than being caught in the surf line and battered to pieces. When this happens, rigging is the best means of removal. Vessels with a great amount of deadrise may come off a reef very nicely on their sides although it may be wise to protect them from damage with plywood or masonite sheets. These can be nailed in place on wood hulls or lashed in

SEAMANSHIP, RIGGING AND BEACH GEAR

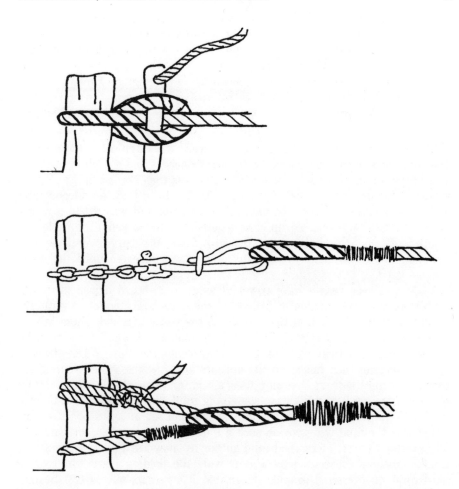

Fig. 25. Quick-release methods for casting off lines under a heavy strain. (Top) Toggle pin. (Center) Pelican hook. (Bottom) Strap and eye release.

Fig. 26. A landing craft refloated with skids and rollers.

place on a fiberglass hull. On other occasions it may be better to jack up a vessel into an upright position. A few heavy-duty oil drums filled with water rolled under the turn of the bilge and chocked with rocks may help keep them upright while they are being pulled off.

With adequate gear and a sound knowledge of rigging and time, practically any vessel can be refloated if it is not too seriously damaged and if it is economically feasible. The trick is not to cause more damage during the process of salving.

In the past beach gear for heavy salvage work was often set up on a barge. The barge was anchored securely with heavy anchors and a cable was passed aboard the grounded vessel. The anchor cable was then heaved up by a heavy wire fall led down the deck of the barge and attached to the anchor cable with a carpenter's stopper. The hauling part of this fall was led to a winch. When the falls were heaved up close together ("two-blocked") the cable was stopped off and the falls were stretched down the barge's deck with a downhaul led to the same winch. The process was repeated until the vessel was refloated.

Modern salvage vessels have powerful winches on board driven by torque converters that are capable of delivering the necessary amount of pull. The limitation is the strength of the fittings on the vessel aground. These salvage tugs usually spot a couple of Eells anchors that are especially designed for salvage work and then pay out the tow cable to the ship. When the ship moves, they can then heave up the anchors and tow the vessel clear of the shoal. The light salvage operator uses a similar principle. In order for his beach gear to work, it must have something solid to work against. If there is mud bottom, a number of drums open at one end set in series on chain and jetted into the mud will provide such a mooring inexpensively. In sand it is often better to fabricate a good sand anchor as shown in Fig. 2. A number of smaller anchors in series may also provide the holding power. In rock or hard coral an oversized grapple, fabricated from scrap, may serve the purpose. Usually a little diving around the area will indicate the best method if the salvor does not know it already.

A typical light salvage beach gear operation is illustrated in Fig. 27. The hull is strapped and a six-part tackle is connected to the wire; the hauling part of the fall is heaved by a chain fall or come-along, but even the boat's anchor winch or perhaps a sailing vessel's sheet winches can heave up on another tackle, compounding the force of the first tackle. In moderately smooth bottom, log rollers, round fence posts, or pieces of 4" pipe placed under the keel might help get the vessel moving on sand bottom. Perhaps a couple of heavy wide planks placed under the rollers may be needed for support. The vessel can usually be levered up enough or jacks can be used to pass the rollers under the keel. Sometimes wedges might also be driven in under the keel to lift the vessel. The salvor has many options and with a little experience, he will be able to determine which method may best suit his purpose.

SEAMANSHIP, RIGGING AND BEACH GEAR

Regular screw jacks lifting upward can raise the entire vessel and will trip as the vessel is pulled seaward. The process can be repeated as needed. Jacking against a vessel's stem can be effective, but is best used in conjunction with an anchor placed to seaward to prevent the vessel being beached again by waves.

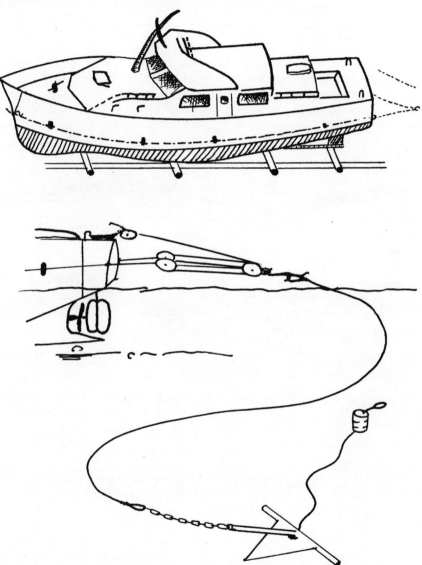

Fig. 27. (Top) Vessel rigged on rollers and beach gear setup. (Bottom) Layout of the beach gear.

CHAPTER 11

Rigging for a Tow

There are basically two methods of towing that salvors use—"alongside" and "hawser" towing. The former is good for shifting another boat or raft around in confined waters, but is not ordinarily suitable for open water work, as the vessels are in contact with each other. Hawser towing is used for both inside and outside work. A short hawser or towline is used in protected or confined waters, while a good long towline is indicated for offshore work. (Fig. 28.)

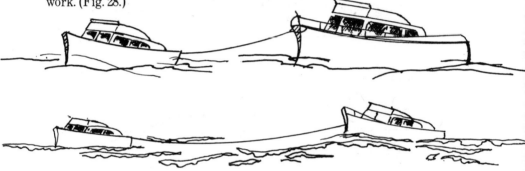

Fig. 28. (Top) Towing on a short hawser in a protected area. (Bottom) Towing on long hawser in the open sea.

It does not necessarily require a great deal of power to tow. A dinghy with a 5 hp outboard in a protected area can easily handle a 50′ yacht in the hands of a capable operator. A good towboatman once stated the essence of towing: "It's more important to have the horsepower in the wheelhouse than in the engine room."

We will deal with alongside towing first. Among commercial towboatmen this is called "on the hip." (Fig. 29.) The towing vessel must first of all have some adequate cleats to make lines fast to and some good lines. A heavy-duty thwart or ringbolt can be used in a pinch, but if the boat is going to see much of this service, it should be fitted out properly. Slack lines are a nuisance when towing "on the hip" and ringbolts and thwarts make it difficult to keep the lines tight. Next, the towboat should have good fenders to save damage to both boats, especially when some idiot in a big power yacht whizzes by to see what you are doing. This will serve as a reminder, too, to have all the cleats well fastened with bolts, not screws.

RIGGING FOR A TOW
53

The towboat should make up alongside with three lines—headline, spring line and stern line, as far aft as is practical. Actually you can make up on the bow just as well as the stern, wherever it is more convenient, but be sure that the tug's stern is far enough behind or outboard of the stem or stern of the towed vessel to give good steering power. If possible, the tug's center line should be slightly inclined inward towards the center line of the other boat. This generally affords better steerage.

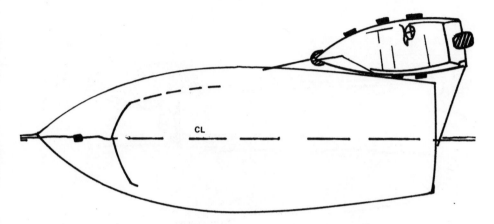

Fig. 29. Towing "on the hip" diagram shows a typical make-up.

Try to make your lines on the other boat fast to deck fittings made for this purpose but many pleasure boats have only limited numbers of cleats, so stanchions, etc., occasionally have to be used. Sailboats usually have enough cleats and bitts and the rigging is usually sturdy enough to take a moderate strain. Keep an anchor handy and make sure the towed vessel's helm is midship. If you are going to berth the vessel on what may well be your blind side, have a friend on deck give you hand signals and handle the dock lines for you.

Examination of a commercial tug that tows barges astern will provide a good idea of how this operation is carried out. (Fig. 30.) Hawser tugs will have a good set of tow bitts located well forward of the rudder, usually about 1/3 of the vessel's length from the stern. This permits them to pivot or turn freely when they have a barge in tow astern on a short hawser. If the hawser were made fast right on the stern, the tug would have difficulty in maneuvering with a large barge in tow. When a tug is towing at sea, it frequently has tow pins in the stern that keep the hawser midship. With a long hawser it works well and the tug can steer effectively. If a salvor's workboat is going to see much duty as a tug, a suitable set of bitts installed or some other arrangement should be made to permit the operator to handle the boat well. A temporary rig can be used with a line made fast on either quarter passing through the thimble of the tow hawser, but this should only be used in the

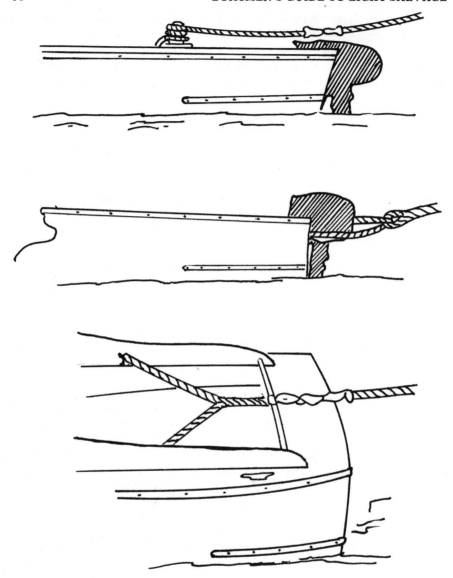

Fig. 30. Different rigs for towing astern. (Top) A small boat with a tow bitt rigged on a thwart. (Center) Method of towing with a bridle passed through a thimble on the towline. (Bottom) Cockpit of a charter fishing boat that occasionally handles tows. Location of bridle ahead of rudder gives good control.

RIGGING FOR A TOW

absence of some better arrangement. A friend of mine installed two heavy eyebolts in the cockpit of his charter fisherman. This allowed him to rig a bridle inboard of his rudders and his boat handled tows very well rigged this way. In a workboat a similar setup might do, but a set of tow bitts could be set up either as a permanent or removable fixture for this purpose.

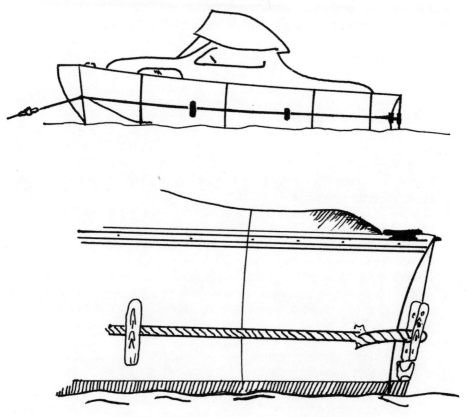

Fig. 31. Method of strapping hull for an offshore tow.

The vessel or object to be towed deserves some attention as well. Many yachts have poorly fastened bitts or cleats forward. If they seem likely to carry away under the strain of towing, it is sometimes better to strap the hull; that is, pass a line all around the hull and secure it in place with lashings and cleats if possible. (Fig. 31.) Chafing gear should be used at all corners to avoid having it part there. The towline is then connected to this strap. Most sailboats or auxiliaries can be towed by securing the towline to the mast or at least reinforcing the bitts by passing an additional lashing from the bitts to the mast.

Nylon makes the best towline, but other fiber lines are better for making fast alongside as nylon is quite elastic. It is not unusual to attach a set of light wire bridles to the towline just as tugs do—the eye of the bridles should be covered with garden hose or else served with marlin to avoid damaging the bitts—the bridles will normally pass through the vessel's chocks. If bridles are not used, then chafing gear of heavy canvas, burlap, or split heavy-duty hose should be attached at any wear point on the towline to avoid having it part.

A 3/8 polypropylene heaving line approximately 100' long is handy to have for passing a towline to a vessel adrift or aground, especially if the sea is rough.

If you intend to try to pull a vessel aground off the beach get your own anchor down well to weather if it is a lee shore and then take a strain on the towline. Small boats may not respond as well to the helm in a heavy surge and the anchor will keep their head to sea while they are pulling on the beached vessel.

Towing operations like everything else at sea require only common sense and some experience to achieve success.

Here are a few suggestions:

1. Secure all shackle pins in tow gear with a seizing.

2. Stream enough towline when the weather is rough. This means enough towline to keep the bight in the water at all times.

3. Stay out of inlets during stormy weather when heavy seas are breaking. (This is good advice whether you have a tow or not.)

4. Take your tow to the nearest safe port. This may entail a longer voyage to leeward in favor of making a shorter but more time-consuming voyage to weather.

5. Put out plenty of chafing gear.

6. A tow that yaws heavily may be more effectively controlled with a heavy length of line trailing astern.

CHAPTER 12

Moving Material Under Water

At times the salvor is faced with the problem of moving quite a bit of material. This may be the accumulated mud, sand and debris that tend to collect in the hull of a vessel that has been sunk for a while, or he may wish to uncover an object that has been buried in the sand or mud. He may also wish to clear away the sand or mud from around a vessel aground in order to refloat it, or jet his ground tackle into the bottom in order to use the beach gear. In each instance there is equipment that can serve his purpose.

There are basically four techniques that are used in operations of this nature:
1. Submerged pumping.
2. Airlifting.
3. Jetting.
4. Scouring.

Open boats that have been under water for a while often fill up with silt. Some of them may be almost completely buried in soft bottom, especially if there is any wave action or current. It may not be necessary but it is often convenient to clear away all of this material before attempting to raise them. The best type of pump for removing the material that has accumulated inside the hull is the diaphragm type of "trash" pump used by plumbers for swamping out septic tanks. These will handle semisolid material and miscellaneous bolts, nuts and washers without clogging.

Airlifting is another method for removing sand and silt and odd-sized material from a site. Treasure hunters use air lifts a lot. Since the air lift is not a pump and does not have impellers to clog and get damaged, the method is popular with diamond and gold divers and commercial salvors. It can suck up any size object that will pass up the pipe. It works on the principle of pressure differential and requires a minimum of about one atmosphere to work well (a depth of about 35'). After that, the deeper it is used, the better it works. Several designs are shown in Fig. 32. The simplest is just an air hose supplying air to a large diameter pipe. Four-inch diameter pipe would be about the minimum practical size for salvage. Diamond divers may use smaller ones for portability. An air lift can be constructed from heavy duty PVC pipe or aluminum which is much lighter than steel in the same sizes, but steel will last longer. The volume of air required is variable, but a 4" pipe will require 30 to 50 cubic feet per minute; larger pipes require proportionately more air. There should be valves to regulate the flow and a shutoff valve for the diver operating the rig. This operation can be dangerous and the diver should keep himself well clear of the intake.

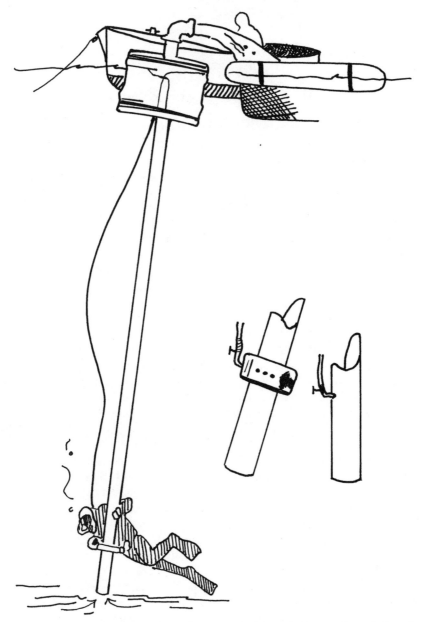

Fig. 32. Diver using air lift which discharges into wire mesh hung under raft in order to recover artifacts.

MOVING MATERIAL UNDER WATER

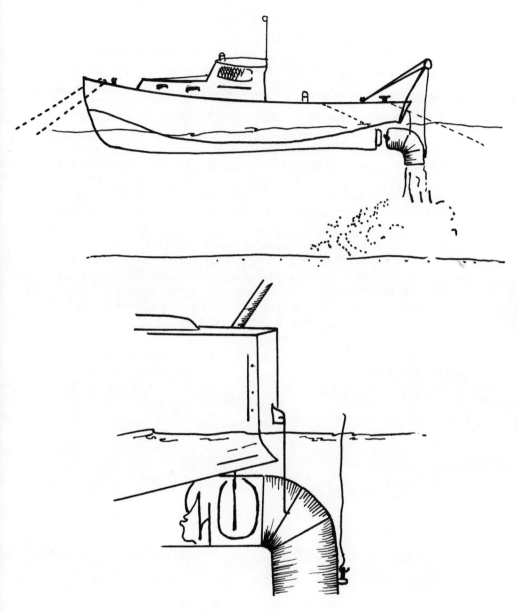

Fig. 33. A boat rigged for scouring. The scouring jet can be stowed easily when not in service.

Jetting is almost self-explanatory. A length of hose with a nozzle is attached to a 2" or 3" centrifugal pump. These develop plenty of pressure to blow away mud and sand. They are also fine for jetting under a boat that has grounded in thick soft mud or sand and with time you might even dredge a channel toward deep water. It will at least clear away around the hull. With wooden boats it is good to avoid directing the jet too much toward the hull as it may dislodge a lot of caulking, particularly in an older boat.

Scouring is just another form of jetting, but it does an excellent job. A thin wall pipe of larger diameter than your boat's propeller is fabricated into a scouring jet so that it directs the propeller flow directly downward; in shallower water, say 10' to 25' depths, the propeller wash from a 200 hp engine will really stir up things and clear away a lot of sand or mud. It is a good idea to mark off the area with buoys and anchor the boat with three or four anchors in order to move its propeller wash around. A little current helps to carry the suspended silt away so that none will settle back again where it was before. It usually takes a little while for the water to clear enough for the effect to be seen. I watched someone clear off a marine railway that had been silted in for years, and it only took him a couple of days to do so. The shipyard owner was amazed as this type of operation had never occurred to him, and he had anticipated spending several weeks of jetting with a small pump. If a boat is going to use this scouring nozzle it is best to rig it on brackets so that it can be raised and lowered as needed. This is illustrated in Fig. 33.

CHAPTER 13

Miscellaneous and Extraneous

We live in an automotive age, yet in spite of this, many or perhaps most of us are incapable of making even minimal repairs to our cars. Garages are located at frequent intervals and wreckers are available to tow us whenever the failure is beyond our limited ability to make the necessary repairs. Many cars travel highways without a flashlight, a pair of pliers or even a screwdriver. We feel secure. Our roadways are well traveled and help is available at the nearest telephone. It was not always like this; I can remember when a set of tools and a towrope were standard equipment in most cars.

The sea is a broader highway than any car will travel, but many of us, secure in our ignorance, voyage there as blithely unconcerned as most motorists ashore. The sea is a capricious mistress, and the ocean's bottoms are littered with the wreckage of vessels whose masters' confidence stemmed from a fatal innocence.

Those who set out to sea should prepare to contend with unexpected problems. Big ships do; they hold regular fire and boat drills and every book on seamanship deals with emergency situations. The yachtsman should take just as many precautions for all too often the assistance of another vessel or the Coast Guard comes too late.

The salvor is often the ultimate beneficiary of a yachtsman's carelessness. But if the yachtsman had applied in advance some of the techniques that the salvor must use later, he might have saved himself some grief and perhaps his life or the lives of those he loves.

I've set out to provide some useful information for people whose purposes are somewhat opposed. The salvor's is to make a living from the accident and misfortune that befall others. The boatman's purpose is to enjoy the sea and avoid the consequences of negligence.

Salvage and safety are unwieldy subjects; ideas and odd bits keep cropping up in a disorderly fashion similar to leftovers in the refrigerator. Included is a list (a sort of hash or stew) of assorted odd bits of information, not in any order of importance but as they occur to me.

    1. Heavy objects may be carried best slung **under** a small boat. Sometimes rigging a pipe through the hull makes it convenient to lift them.

    2. Large sheets of light plywood can often cover large curved areas by laying it on a bias, rather than fore and aft or up and down.

    3. Channels into reefy areas are often best explored by a diver, who can buoy them with large plastic bottles. Fluorescent red and green tape can be attached to indicate the appropriate side of the channel when it is illuminated by a spotlight at night. It might help to number them too.

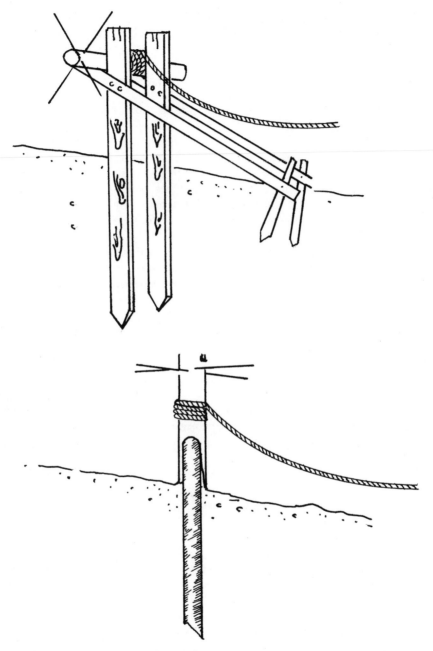

Fig. 34. Hand winches made from timbers or pipe can be used for working on beaches.

4. Small high-intensity flashing lights used for life rings might be more conspicuous hoisted aloft on the mast of a sailing vessel when assistance is required than attached to a life ring that is on board. A powerboat might be able to rig it to its radio antenna or fishing outrigger to increase its range of visibility.

5. A small triangular file is better than a stone for putting a quick sharp edge on a knife for cutting lines.

6. Chain binders are helpful for securing or lashing objects.

7. Windlasses and capstans can be erected easily out of pipe and timbers as shown in Fig. 34.

8. Pneumatic and hydraulic tools can be used under water. Hydraulic tools are better since they don't require as much power to drive them as air. Electrical tools are a "no-no."

9. Mouse the hooks on block and tackles so they won't come adrift when the fall is slacked.

10. Saw wires with finger loops are handy for cutting off propeller shafts and rudders, etc. Sometimes they can be operated by pulling on lanyards attached to the finger rings.

11. A hook turned in the end of a piece of reinforcing rod will help pass lines under a boat without crawling under it.

12. Silicon sealer helps seal dry seams above the waterline. This can be applied with a putty gun.

13. C-clamps are handy for holding things in place until they can be fastened.

14. The salvor should be prepared to expend some effort on preserving the machinery and the electronic equipment on salvaged vessels, especially if they have been under water. The time involved will usually be well compensated since the salvor's award is based on a percentage of the vessel's value. If the main engine and all the other equipment is a total loss, this has a definite effect on the value of the vessel.

Chapter 14

Some Legal Aspects of Salvage

The nature of salvage work encourages litigation. Agreements made in haste and under pressure of circumstance are easily forgotten. Salvors, too, are sometimes greedy. The court becomes the natural arena where these disputes are settled. Nothing is new about this situation; some legal scholars speculate that the basis of admiralty law as it exists today may extend as far back as 900 B.C. when the Island of Rhodes is believed to have established a code of law dealing with maritime affairs. Be that as it may, there is a certainty that eventually any active salvor will encounter occasions when he must look to the courts in order to receive his compensation. The boatman himself fares best there, too, when pressed by a greedy or unscrupulous salvor. The law is a double-edged sword, and can and does cut in both directions.

The law in a court of Admiralty differs markedly from the common law ashore, and is more international in character. There are several criteria that apply which determine whether an act can be legitimately classisfied as **salvage**. If these criteria are met, any legal action that follows is considered **maritime** in nature and falls within the jurisdiction of the federal court.

1. It must take place on navigable waters.
2. It must be a volunteer action.
3. It must be successful.

Navigable waters are those upon which international or interstate transportation is carried on, or waters connecting to them. A lake or river whose boundaries fall entirely within one state is not considered navigable by the courts, while another body of water whose boundaries fall within two or more states is considered navigable.

The term **volunteer action** is sometimes confusing. It means that the salvor must be under no compulsion or obligation to act. This omits salvage liens in favor of members of the armed forces or professional crews employed aboard salvage vessels. There are usually "riders" attached to the articles of such vessels. But the owners of the vessel may seek remedy in Admiralty court.

A salvor's award is predicated to a large extent on the value of the property saved. Ordinarily, unless circumstances beyond his control intervene, the salvor will have no claim against the property for his efforts, unless he is successful. The standard salvage contract usually stipulates this and in the trade this is what the term "No Cure, No Pay" means.

Salvages normally fall into two basic categories. "Pure salvage" and "contract salvage." The common instances of pure salvage occur when an individual undertakes without prior agreement to salvage a vessel if the vessel is manned, and his services are accepted, he then has a salvor's lien against the vessel provided his efforts were successful. If he cannot then arrive at a mutually-agreed-upon fee for his services, he is dependent upon the court's discretion with regard to the amount of his award. In other instances he may act where there is no one aboard to accept or reject his services. Upon successful completion of his salvage he is entitled to a salvor's fee and if he cannot come to an agreement with the owner, he must again seek his remedy in Admiralty court.

Contract salvage, on the other hand, usually avoids the litigatory process by agreeing in advance to the terms, or signing an agreement where both parties submit to arbitration. Lloyd's standard salvage form is a good example of this. When both parties agree to this "No Cure, No Pay" contract and the salvor is successful, the owner of the vessel posts a bond, and the award is made by a panel of experts who act as arbiters. This avoids the costly legal expenses and delays inevitable in the judicial process.

The salvor does not own the property he has salvaged, but has a right of possession to it until such a time as the owner either settles his claim or posts a bond pending an award in Admiralty court. In the latter case, the salvor normally files a maritime lien for salvage against the property and the Federal Marshal takes possession until the bond is posted. If no bond is posted, the Marshal will retain possession until the matter is settled in court.

Here are a few tips for both the salvor and the boatman:

1. Keep an accurate log. Courts make their decisions regarding the award based on sound information. A good log helps clarify matters. The factors involved are risks to the property saved and to the salvor, and danger to his own equipment; the quality of seamanship required and the promptness with which the services were tendered; the amount of effort involved and material consumed in the process.

2. Salvage the equipment, but resist any temptation to make off with portable items. Admiralty courts are generous to salvors, but frown on looting or theft. Curb your crew if they are light-fingered.

3. Engage a good attorney familiar with Admiralty proceedings and invest in a few books for yourself. *Law of Tug, Tow and Pilotage*, by Alex L. Parks (Cornell Maritime Press, Inc., Centreville, Md.); and *Your Boat and the Law*, by Martin J. Norris (Lawyers Cooperative Publishing Co., Rochester, N.Y.) should be in every salvor's library.

4. Bill for salvage services only, as it is a high priority lien, ahead of all others with the exception of seamen's wages. Get a signed contract beforehand if you can.

SAMPLE CONTRACT

I, _____, Master/Owner of the _____ agree to accept the services of _____ Salvage Co. on a "No Cure, No Pay" basis in the sum of _____ ($) _____ to be paid in hand upon successful completion of salvage. Salvor retains right to possession of said vessel until account is settled in full or bond of sufficient amount is posted.

                              Signed _____

                              Signed _____

Witnessed _____

Witnessed _____

Chapter 15

Getting Started in Light Salvage

In the salvage business as in any other, the hardest part is getting started. Another factor worth considering is that salvage work tends to be an inconsistent proposition at best. Most full-time salvors built their business from what was initially a part-time operation. In other words, don't give up a good job in hopes of becoming an immediate success in the salvage game.

There was one operation in particular that typifies a good small light salvage operation. A retired Navy Chief Petty Officer managed a small marina. His duties were light and his schedule was flexible, so he fitted out a surplus LCPL hull (a Higgins type utility boat) for towing. He was located in a popular boating area and knew that on weekends and holidays the Coast Guard was working full-time attending to the needs of pleasure boats in distress. He had three things in his favor:

1. He was an experienced boatman.
2. He had adequate equipment (a 36' boat with a 200 hp diesel engine).
3. He had a modest but steady fixed income.

He was aware that the Coast Guard was overworked as it was, and also that they are reluctant to tow in vessels not in immediate danger of sinking or going ashore if commercial assistance is readily available. He made himself available and advised the Coast Guard of his readiness to provide assistance to other vessels and gave them a telephone number where he could be reached.

He also monitored the VHF more or less continuously for distress calls. He charged reasonable rates for his services (night time rates were higher than daytime rates). He also maintained friendly relations with the local charter boat fleet and they frequently advised him of the location of vessels in trouble. The charter boats would help anyone in a genuine emergency, but in the case of a simple breakdown, they referred the business to him because assisting these vessels interfered with their obligation to their own charter party.

He built a successful business with a limited investment and provides a public service as well. He relieves the Coast Guard of a portion of their burden which leaves them free to attend to the serious cases that do require their attention. He has to be fairly forceful at times and insists that a salvage agreement be signed beforehand as this gives him a high priority lien against the vessel in case the owner refuses to pay afterwards. It seems that there are always a few deadbeats who own expensive yachts that expect the Coast Guard to provide free assistance whenever they get into trouble. The Coast Guard's services are first rate, but they are not free. We all collectively pay for them. It is rather refreshing when the careless boatmen are required to foot their own bills for a change.

Another source of business for salvors are the marine insurance companies. They look to the salvor to protect their interest, and are usually generous in settling his claim. There is an excellent lever that a salvor can use in getting his services accepted, if the master or owner of a vessel refuses his assistance, the underwriter has grounds to cancel the vessel's insurance. Insurance companies sometimes engage salvors to recover vessels that are insured for TLO (total loss only). It is cheaper to pay the salvor and return the vessel to the owner than it is to pay off the claim for the loss of the vessel. This is most often done in cases where the loss may have seemed a bit suspicious.

Some salvage operators follow in the wake of hurricanes or severe storms. They may move their operation temporarily to the disaster area as there is usually an abundance of work for everyone. Other salvors speculate on damaged or sunken vessels, buying them "as is, where is" from the insurance companies and make their profit by repairing and reselling the vessel later.

In any event, the beginning salvor must "pay his dues." A lot of the business is built on word of mouth. He must circulate around and get to know as many people as he can in the boating industry. This would include yacht brokers, marina and boatyard operators, etc., and take the small tough jobs that no one else wants. If he does good work he will build a reputation that will attract more business. If he is careless or slipshod, that news will get around, too. Salvage is a tough, competitive and very interesting business. With skill and luck and lots of hard work it can be successful. Good Luck.

INDEX

Admiralty law and courts, 64-65
   books about, 65
Air injection, 23, 25, 29
Air lifting, 57, 58
Anchor cable, 50
Anchor hawk, 35
Anchor(s), 2, 8-9, 11, 13, 33, 37, 53, 56, 60

Barges
   salvaging, 2, 23, 25, 29
   use in salvaging, 50
Beach gear, 13, 15-16, 50-51
Blocks and falls, 44, 47-48, 50
Boat for salvage use, 6, 7, 31, 53-54
Boatman, 1, 3, 20, 32, 67
Boatmanship, 4
Bridles, 54
Buoying an object, 37
Buoys, 11, 33, 60

Carpenter's stoppers, 39, 42, 50
Carpenter's tools used in salvage, 6
Chafing gear, 31, 55-56
Chain falls, 9, 44, 48, 50
Chain stopper, 39, 43
Coast Guard, 1, 61, 67
Come alongs, 9, 16, 44, 48, 50
Compressed air for raising submerged vessels, 23, 29
Compressors, 4, 9, 25

Dewatering, 14, 17, 20
Diving, 1, 3, 5-6, 33-37, 57-58, 61
Dragging devices, 35
Drifting vessels, 10, 12
   location of, 31-32
Drop line, 33, 34

Eells anchor, 50
Equipment
   for rescue towing, 31
   for salvage, 6, 9

Fairleads, 44
Fathoms (*see* Working lines)
Fire, 10
Flemish (Dutch) eye splice, 44, 46
Flotation, 23, 25-27, 37, 48
   examples of external, internal, 28
   gear, 26

Grapple anchor, 8, 50
Grapple "rake," 35
"Ground effect" factor, 13
Grounding, 10, 12

Hawser(s), 4, 38, 43, 52-53
"Heaving down," 15
Hitches, 39-41
Hydraulic tools, 63

Jetting, 60

Knots, 39-41

Ladder search pattern, 32
Landing craft, 12, 13, 49
Lashings, 44, 48, 55
"Law of the Sea," 1
Leaks
　common sources of, 20
　repair of, (*see* Patching)
Life boats
　use in raising submerged vessels, 30
Lifting a submerged vessel, 30
Light marine salvage operations, 2-3, 14, 17, 50
LORAN lines, 37

Manifold(s), 25, 27
Marine insurance companies, 2, 68
Marlinspike (*see* Seamanship)
Mudhook, 16

"No Cure, No Pay" contract, 64-65

Oil drums
　use in refloating vessels, 13, 16

Patching, 5
　emergency and semipermanent, 17
　kinds, 17-22
　fiberglass hulls, 20
Pneumatic tools, 63
Power boats
　use of, 16, 63
Pumping
　cleaning out submerged vessels, 57

Pumping (*continued*)
　limitations on use, 23
　raising submerged vessels, 23-24
Pumps, 4, 9, 17, 22, 60
　portable, 16, 17, 31
　trash, 57

Quick release methods for casting off lines, 48-49

Range marks, 36, 37
Refloating vessels, 12-16
Rescue towing, 31-32
　necessary gear, 31
Rigging, 4-5
　temporary for towing, 53-54
　work, 48-50
Rollers, 49-51
Rope stopper, 39

Safety precautions, 61-63
Sailboat(s), 23
　refloating, 12-13
　salvaging of, 15, 28
Salvage business
　getting started in, 67-68
Salvage contract(s), 64-67
　sample, 66
Salvage operations, 1-3, 10-15, 31-32
　legal aspects, 64-65
　underwater, 33-37
　vessels for, 50
　(*see also* Light marine salvage operations)
Salvor(s), 1, 11, 61
　abilities necessary, 4-5, 38, 40
　commercial, 13, 14, 33, 57
　compensation for, 2
　equipment for, 6, 9
　light, 14, 17
　risks, 3

INDEX

Salvor(s) *(continued)*
  work of, 13, 14-16, 20, 23, 25, 27, 31-32, 33, 44, 48, 53, 63
Sand or mudhook anchors, 8, 50
Scouring, 59-60
Screw jacks, 51
SCUBA gear, 3, 6
Seamanship, 1, 4, 38, 61
Skids, 48, 49
Slings, 43-45
Speculation in distressed vessels, 68
"Standpipes," 25, 27, 29
Stern anchor, 13
Stoppers, 39, 42-43
Straps, 43-44
Submerged objects
  salvage of, 22 *(see also* Salvage operations)
Sunken vessels, 10

Tackles, 47-48
  power ratios, 47 *(see also* Blocks and falls)
Taking marks, 4, 11, 36-37
Tide tables
  use of, 16

Towing, 15-16, 52-56
  bitts, 54, 55
  "on the hip" (alongside), 52-53
  rigs for, 54
  sailboats, 53
  strapping a hull for, 55
  tips for success with, 56
Towlines, 31, 34, 56
  (see also Hawsers)
Tug boat, 53

Underwater
  locating of an object, 37
  search, 33-35

Water Plug (Quick-set Water Plug), 18
Winch(es), 44, 48, 50, 61, 62
Windlasses and capstans, 62, 63
Workboat, 7, 53
Working lines, 38-39
Work raft, 7, 9
Work scow, 7, 9

Yachts
  salvaging of, 12-13, 16, 23, 28

ANF 627.703 REID   02631522
Boatmen's guide to light salvage /   SPL
Reid, George H., 1924-
Cornell Maritime Pres 1979.
3 3610 00436 3558